Biotechnology

Theory and Techniques

Volume I

The Jones and Bartlett Series in Biology

AIDS Smartbook, Kopec/Wood/Bennett

Anatomy and Physiology: An Easy Learner, Sloane

Aquatic Entomology, McCafferty/Provonsha

Biology, Ethics, and the Origins of Life, Rolston

Biochemistry, Abeles/Frey/Jencks

Biology: Investigating Life on Earth, Second Edition, Avila

The Biology of AIDS, Third Edition, Fan/Conner/Villarreal

Biology: Investigating Life on Earth, Second Edition, Avila

Biotechnology, Theory and Techniques, Volume I, Chirikjian

Biotechnology, Theory and Techniques, Volume II, Chirikjian

The Cancer Book, Cooper

Cell Biology: Organelle Structure and Function, Sadava

Cells: Principles of Molecular Structure and Function, Prescott

Creative Evolution?!, Campbell/Schopf

Early Life, Margulis

Electron Microscopy, Bozzola/Russell

Elements of Human Cancer, Cooper

Essentials of Molecular Biology, Second Edition, Freifelder/Malacinski

Essentials of Neurochemistry, Wild/Benzel

Evolution, Second Edition, Strickberger

Experimental Research Notebook, Jones and Bartlett Publishers

Experimental Techniques in Bacterial Genetics, Maloy

Genetics, Third Edition, Hartl

Genetics of Populations, Hedrick

The Global Environment, ReVelle/ReVelle

Grant Application Writer's Handbook, Reif-Lehrer

Handbook of Protoctista, Margulis/Corliss/Melkonian/Chapman

Human Anatomy and Physiology Coloring Workbook and Study Guide, Anderson

Human Biology, Donald J. Farish

Human Genetics: The Molecular Revolution, McConkey

The Illustrated Glossary of Protoctista, Margulis/McKhann/Olendzenski

Major Events in the History of Life, Schopf

Medical Biochemistry, Bhagavan

Methods for Cloning and Analysis of Eukaryotic Genes, Bothwell/Yancopoulos/Alt

Microbial Genetics, Second Edition, Maloy/Cronan/Freifelder

Molecular Biology, Second Edition, Freifelder

Oncogenes, Second Edition, Cooper

100 Years Exploring Life, 1888-1988, The Marine Biological Laboratory at Woods Hole, Maienschein

The Origin and Evolution of Humans and Humanness, Rasmussen

Origins of Life: The Central Concepts, Deamer/Fleischaker

Plants, Genes, and Agriculture, Chrispeels/Sadava

Population Biology, Hedrick

Statistics: An Interactive Text for the Health and Life Sciences, Krishnamurty/Kasovia-Schmitt/Ostroff

Vertebrates: A Laboratory Text, Wessels

Biotechnology
Theory and Techniques
Volume I

Plant Biotechnology
Animal Cell Culture
Immunobiotechnology

This book was written with support, in part,
from the National Science Foundation

Editor

Jack G. Chirikjian
Georgetown University School of Medicine, Washington D. C.
Chairman, EDVOTEK, Inc., Rockville, Maryland

Associate Editors

Edward Kisailus
Canisius College
Buffalo, New York

Baldwin King
Drew University
Madison, New Jersey

Robert Krasner
Providence College
Providence, Rhode Island

Harley Mortensen
SW Missouri State University
Springfield, Missouri

Managing Editor

Karen Graf
EDVOTEK, Inc.

JONES AND BARTLETT PUBLISHERS
Boston London

Editorial, Sales, and Customer Service Offices

Jones and Bartlett Publishers
One Exeter Plaza
Boston, MA 02116
617-859-3900
1-800-832-0034

Jones and Bartlett Publishers International
7 Melrose Terrace
London W6 7RL
England

ISBN 0-86720-895-3

Printed in the United States of America
99 98 97 96 95 10 9 8 7 6 5 4 3 2 1

CONTRIBUTORS

Jack G. Chirikjian
Georgetown University School of Medicine, Washington, D. C.
Chairman, EDVOTEK, Inc., Rockville, Maryland

Dennis Bogyo
Valdosta State College
Valdosta, Georgia

John Boyle
Cerritos College
Norwalk, California

Audrey Brown
Seattle Central Community College
Seattle, Washington

Jordan Choper
Montgomery College
Takoma Park, Maryland

G. Bruce Collier
Trevigen, Inc.
Gaithersburg, Maryland

Richard Echols
Southern University
Baton Rouge, Louisiana

Mark A. Holland
Salisbury State University
Salisbury, Maryland

Barbara Jones
Southwestern Adventist College
Keene, Texas

Baldwin King
Drew University
Madison, New Jersey

Edward Kisailus
Canisius College
Buffalo, New York

Karen K. Klyczek
University of Wisconsin-River Falls
River Falls, Wisconsin

Robert Krasner
Providence College
Providence, Rhode Island

Malethu T. Mathew
Virginia Union University
Richmond, Virginia

Harley Mortensen
SW Missouri State University
Springfield, Missouri

Ah-Kau Ng
University of Southern Maine
Portland, Maine

Mark Petersen
Blue Mountain Community College
Pendleton, Oregon

E. Robert Powell
Central Oregon Community College
Bend, Oregon

Geraldine Ross
Highline Community College
Des Moines, Washington

Anthony Sena
Northern New Mexico Community College
Espanola, New Mexico

Ellie Skokan
Wichita State University
Wichita, Kansas

Kathy Steinert
Bellevue Community College
Bellevue, Washington

Peter Woodruff
Champlain Regional College
St. Lambert, Quebec

CONTENTS

TO THE INSTRUCTOR

The work presented in the two volumes of Biotechnology Theory and Techniques is the collaborative effort of over twenty undergraduate science faculty. The undergraduate faculty represent various two-year and four-year colleges in several regions of the United States and Canada. The common goal of the faculty contributors was to develop a publication containing unique and flexible laboratory activities that focus on the theory and practice of biotechnology for students.

The experiments presented in both volumes are designed to utilize many resources readily available in most science departments. Care was taken to avoid duplicating research protocols and laboratory experiments that are available in a variety of existing laboratory manuals. We believe that Biotechnology Theory and Techniques is a unique and practical compilation of experiment modules that are safe, accurate, and appropriate for implementation in academic science laboratories.

The topics and experiments included are deemed by the faculty to represent major areas of biotechnology, but by no means address all aspects of the continuously evolving technology. The two volumes are not designed for any specific science course. The strategy is to allow the flexibility for selecting various modules to be implemented in life science courses, and wherever else they are applicable. Alternate approaches are included.

As Editor of the two volumes of Biotechnology Theory and Techniques, my role is primarily that of a catalyst. I am fortunate to be able to assemble a creative and dedicated group of science college faculty whose goal is to enhance a classroom experience for student interested in learning about the science of biotechnology.

TO THE STUDENT

Although the protocols, techniques, and procedures organized in this manual have been developed over years of work by scientific investigators, these laboratories should not be considered as canned, nor will they succeed in every way, every time. They should be looked upon as experiences leading to adventure, challenge, and discovery.

Your instructor may want you to keep a separate duplicate laboratory notebook for your write-ups. The write-up is a very important element in the actual learning of laboratory science. It is precisely in the doing of laboratories such as those contained herein that you learn how to do research and develop a sense of how to approach a problem, seeking and recording data as you proceed. You should not be discouraged if you do not get exact results in each experiment. Always ask yourself: What could have been the reason for the different results? What could I do differently that might lead to the desired outcome?

Always remember that it has taken years of invested time for teachers and researchers to develop the laboratory experiences that you will be attempting with this manual. These exercises are not chosen for ritualistic value, but rather to expose you to the wonderment of the experimental spirit that is the essence of science. Enjoy the adventure.

ACKNOWLEDGMENTS

Thank you to Jones and Bartlett, our publisher, for ensuring the earliest possible publication of our work. We appreciate the marketing wisdom of our sponsoring editor, David Phanco and the efforts of editorial assistant, Deborah Haffner as well as the production editors, Mary Cervantes-Sanger and Nadine Fitzwilliam.

I am also grateful to members of the group for their individual and collective contributions. A special thanks is extended to the staff at EDVOTEK who contributed to this work, and to Edward Kisailus, Baldwin King, Robert Krasner, and Harlye Mortensen, the four associate editors of the project. Special recognition goes to Karen Graf, managing editor, whose energy and dedication has made the completion of the project possible. Recognition is also extended to Kathy Gilbert, editorial assistant, for her contribution to the presentation of the experiment modules. A final special note of appreciation goes to the National Science Foundation, which provided support for initiating the project.

Jack G. Chirikjian, Ph.D.
Professor of Biotechnology and Molecular Biology
Georgetown University School of Medicine, Washington D.C.
Chairman, EDVOTEK, Inc. Rockville, Maryland

UNIT 1

Plant Cell Culture

UNIT 1: PLANT CELL CULTURE

MODULE 1: ESTABLISHING A PLANT CELL CULTURE

Mark Holland

* Introduction
* Safety Guidelines
* Experimental Outline
* Materials
* Pre-lab Preparation
* Method
* Results

General Unit Introduction

This unit consists of three modules which introduce some of the basic techniques of plant cell culture. In Module 1, students will establish cell cultures from a variety of plant materials, learn about the components included in plant culture media and practice sterile technique. Module 2 presents a protocol for releasing protoplasts from plant tissue and includes a simple experiment demonstrating protoplast fusion as an example of plant cell manipulation. In Module 3, students will set up a plant regeneration experiment and learn about the effects of plant growth regulators (plant hormones) on plant cells in culture.

The modules in this unit require only modest equipment and the media can be purchased ready-made, if desired. Plant cell culture techniques offer a good contrast to animal cell culture techniques described in another unit of this series. Plant cell cultures require somewhat simpler facilities and require less attention than do animal cultures, but individual experiments with plant cells may take longer to complete. Of course, the outstanding result for many students is the amazing ability of plant cultures to regenerate whole plants.

Introduction

When a plant is wounded, cells at the wound site proliferate into an undifferentiated mass called callus tissue. This tissue can be thought of as a kind of scar tissue and, in nature, callus formation seals the wound against attack by pathogenic bacteria and fungi. Callus is probably most familiar to you as the swollen, knot-like growth seen on trees at the site of limb removal, but herbaceous plants as well as woody plants produce callus. Plant cell culture exploits this natural wound response of plants by encouraging callus growth on small pieces of excised tissue or "explants".

On artificial media, callus can be maintained in a state of persistent, undifferentiated growth. By changing the concentration of plant hormones in the medium, however, the callus of many plant species can be made to redifferentiate into whole plants. The ability of single cells to regenerate whole plants is referred to as totipotency.

The commercial applications of plant tissue culture techniques are quite important. Regeneration of plants from cell culture offers a practical strategy for plant cloning, since all regenerants from a culture should be genetically identical. For commercially valuable plants that are difficult, costly or inefficient to propagate by cuttings or other asexual means, cell culture sometimes offers the only practical means of propagation. Most hybrid orchids, for example, are propagated today by the tissue culture method of "meristemming" or "mericloning". Plant cultures are also being investigated as sources of valuable plant products like drugs, flavors and fragrances.

Many strategies for genetic engineering of plants rely on plant tissue culture. Plant cells in culture can be genetically transformed by a number of techniques. Explants can be transformed directly by *Agrobacterium tumefaciens* or by bombardment with DNA-coated particles from a particle gun. Protoplasts (plant cells with their cell walls removed) are the targets of microinjection and electroporation (DNA uptake mediated by an electric field). Plant cell culture is central to these techniques in that cell culture allows transformants to proliferate and, sometimes, regenerate into genetically identical clones.

While many formulations for plant tissue culture media have been developed, all contain the same basic types of ingredients. These are:

1. Inorganic salts
   Macro and micronutrients
2. Vitamins
3. An organic carbon source

Introduction, continued

In addition to these components, most formulations contain plant growth regulators (plant hormones), auxin and cytokinin. Callus media are generally solidified by the addition of agar. (Plant cultures are sometimes maintained as a suspension of cells in liquid medium, but this type of culture will not be included in this laboratory experiment.)

The medium used in this experiment, MS medium, is a formulation developed by Murashige and Skoog (1962. Physiologia Plantarum 15:473-497.) MS medium is one of the most commonly used of all plant culture media. The ingredients included in this medium are:

	mg/l
1. Inorganic Salts	
Macronutrients	
Ammonium nitrate	1650
Potassium nitrate	1900
Calcium chloride (anhydrous)	332
Magnesium sulfate (anhydrous)	180.7
Potassium phosphate	170
Micronutrients	
Ethylenediaminetetraacetic acid (EDTA, Disodium Salt)	37.3
Ferrous Sulfate (heptahydrate)	27.8
Manganese sulfate	16.9
Zinc sulfate (heptahydrate)	8.6
Boric acid	6.2
Potassium iodide	0.83
Sodium molybdate (dihydrate)	0.25
Cobalt chloride (hexahydrate)	0.025
Cupric sulfate (pentahydrate)	0.025
2. Vitamins	
Myo-inositol	100.00
Thiamine hydrochloride	0.40
3. Carbon source	
Sucrose	30,000.00

The medium is solidified with 7.5 g/l of agar. The amounts of plant growth regulators added to the medium are variable, depending on whether the culture is to be maintained as callus or made to regenerate whole plants. The medium used in this experiment contains Indole Acetic Acid (IAA, an auxin) at a concentration of 1.0 mg/l. No cytokinins are added. The effects of varying the concentrations of plant growth regulators on the cell culture will be investigated in another module (Plant Cell Culture III: Plant Regeneration and Cloning).

Safety Guidelines

Follow standard laboratory safety practices.

Preparation of media requires an autoclave or pressure cooker. Use of this equipment requires care to avoid serious burns.

An alcohol or gas flame is used to sterilize instruments during the lab exercise. Due care should be used with open flames. Precautions should be taken to avoid igniting hair or clothing.

Experimental Outline

Disinfect plant tissue and prepare explants.

Inoculate medium with plant tissue.

Observe cultures over a period of several weeks.

Materials

Sterile media (this has been prepared prior to class)

Plant material

10% Household bleach solution (prepare fresh)

Sterile Water

70% Ethanol

95% Ethanol

Forceps

Scalpel

Pre-lab Preparation

Tissue culture medium should be prepared in advance of the laboratory period. This can be done several days to a week ahead of time. Prepared media can be stored at room temperature. Plant materials used for the experiment should be purchased or collected shortly before the class meeting.

Preparation of plant material for culture, i.e. surface sterilization, will take approximately 30 minutes. Plan another 30 minutes for novices to prepare explants and inoculate the culture medium. The time required for students to complete the exercise will depend on whether students are waiting for space in a laminar flow hood.

Depending on the length of the laboratory period, you may want to combine this exercise with Module 3: Regeneration and Cloning. Both exercises require that the students initiate cell cultures.

Because results from both Modules 1 and 3 require several weeks of observation after the initial laboratory class period, these exercises are best used early in the semester.

Timetable of Events

In Module 1, students will learn how to establish plant cell cultures from a variety of plant tissues. In accomplishing this, students learn how to prepare plant tissue for culture, practice sterile technique, and learn about the components of plant culture media.

The "bare bones" methods described in the student guide for this module (and for others in the plant tissue culture unit) are adaptable to most lab conditions. Depending on the availability of sterile transfer hoods, specialty glassware, etc., you may want to elaborate on the protocols presented, but very little specialized equipment is actually required for success.

Equipping the Work Area

Plant tissue culture is normally done in a laminar flow hood, a containment hood, or in a transfer cabinet. The laminar flow hood has a box shape with one open side facing the worker. Fans in the hood circulate air through a particle (HEPA) filter which removes airborne contaminants. Air flow in the laminar flow hood is from the back of the box outward past the worker. Sterility in the work area is maintained by this flow of clean air, but can be compromised by introducing contaminated equipment into the hood, by leaning into the hood while working, or by placing non-sterile items between open cultures and the back of the hood.

Pre-lab Preparation, continued

Containment hoods , like laminar flow hoods, depend on air flow to maintain sterility in the work area. Because these hoods are designed to prevent the escape of cultured cells from the work area, however, the opening at the front of the hood is partially closed (generally by a window) and air travels from the top of the cabinet downward to intakes on the work surface. Air does not flow out of the hood into the lab environment without first being filtered to remove contaminants. The protection offered to workers by a containment hood is unnecessary for most plant tissue culture purposes, but the hood is very satisfactory for handling plant cultures.

Transfer cabinets (or closely-related glove boxes) are box-like enclosures, sealed on the top and on three sides, with limited access from the front. Sterility is maintained within the cabinet only to the degree that air movement from outside the box to the inside is restricted. The transfer cabinet is really no more than a shelter against airborne contaminants. When used properly, the transfer cabinet is an economical and effective enclosure for cell culture. If desired, a transfer cabinet can be economically "homemade" from plywood or plastic. Painting the plywood surface with enamel will make it washable. (We have actually tried using a sturdy cardboard box as a transfer cabinet. While this works, it is not recommended for the student lab because of the fire hazard.)

In the absence of a sterile hood or transfer enclosure, plant culture experiments can still be carried out. However, depending on the cleanliness of the lab environment and the sterile technique of the students, losses to fungi and bacteria can be expected. The best results will be obtained if windows are kept closed to restrict air movement and if bench tops are wiped with 70% ethanol before work begins. Perhaps most important of all, students should wash their hands well before doing any culture work.

The equipment required for plant cell culture is modest. For handling cultures, the work station should be outfitted with a Bunsen burner, a covered container of 95% ethanol, forceps (long enough to reach into culture flasks or tubes), and a sharp scalpel. For disinfecting the work area, a wash bottle or spray bottle of 70% ethanol and a supply of clean (preferably autoclaved) wipers or paper towels will suffice. For media preparation, you will need a balance weighing to 0.1 mg, a pH meter, and an autoclave. (Note that pre-prepared, even pre-sterilized media are available commercially. Names and addresses of some of these sources are listed at the end of this guide.) Plant cell cultures can be grown in petri dishes, baby food jars, test tubes, Erlenmeyer flasks, mason jars, etc. The only requirements for a suitable container are sterility (they should be autoclavable or presterilized/disposable) and they must

Pre-lab Preparation, continued

be able to be opened and closed repeatedly without introducing contaminants. Cotton or foam stoppers can be used as closures. Specially-designed caps for tubes, flasks and baby food jars are also available. Ideally, plant callus cultures are incubated at 26-28°C, but they will grow at slightly cooler room temperature. Callus cultures can be maintained either in the light or in the dark. In the classroom situation, a drawer in the lab bench can serve the purpose of incubator. Callus cultures from which plants are being regenerated should be maintained under fluorescent lights with a day length of 16 hours.

Preparation

Obtain plant material: A number of different plants can be tested in this experiment. Students may even want to try some materials of their own. The best material for explants comes from plants raised indoors or in the greenhouse because they seem to be "cleaner". Some vegetable tissues are notable exceptions to this. Some plants to try are: Tobacco stem or leaf, carrot roots, African violet leaf or petiole (this material will be used in Module 3), cauliflower florets. If you want to use vegetables from the grocery, try to purchase these as fresh as possible. Plants other than those listed above will also do very well. Make the selection of explants part of the experiment.

Make media: The recipe for MS medium is written out in the student guide to the laboratory. If you choose to assemble the medium from stock bottles on the lab shelf, it will be easier to prepare a 100 x concentrated stock solution of ingredients listed as micronutrients and keep this stock in the refrigerator. Also, plant growth regulators are more easily handled as stocks. Dissolve IAA in a small amount (few mls) of 1 M NaOH, then dilute to 1.0 mg/ml with distilled water. Dissolve kinetin in a small volume of I M HCl, then dilute to 1.0 mg/ml with distilled water. Just before autoclaving, adjust the pH of the plant medium to 5.8.

Pre-prepared media are available commercially and offer a simple and cost-effective alternative for media preparation. For this experiment, order "MS salts with Minimal Organics". To this powder, you will only have to add sucrose (table sugar from the grocery store will work; add 30 g/l), agar (7.5 g/l), and plant growth regulators (For this experiment, add only IAA; 1 mg/l).

Method

1. Prepare plant material for culture. Select fresh-looking, healthy (not brown, bruised or wilted) plant material. This can be leaf, stem, or root. Cut the tissue into manageable pieces. These should be small enough to fit into a beaker for disinfecting, but large enough so that they can be trimmed after disinfection.

2. Move your work into a sterile transfer hood. If a hood is not available and you are working at the lab bench, wipe the work area with a solution of 70% ethanol before beginning and try to keep your work covered as much as possible. Before proceeding to disinfect your plant tissues, wash your hands thoroughly.

3. Wash the explants in a beaker of distilled water to which you have added a few drops of detergent.

4. Using forceps, transfer the explants to a 70% solution of ethanol for 2 minutes.

5. Again using forceps, transfer the explants to a 10% solution of household bleach for 5-10 minutes. Tender leaf tissue should not be left in the bleach for more than about 5 minutes. Root or stem pieces can be left for longer times.

6. After treatment with bleach, the tissue is considered to be sterile. Care should be taken to avoid re-contaminating it. At your work station should be a covered container of 95% ethanol and a Bunsen burner. Before they are used to handle sterile plant tissues, forceps and scalpels should be dipped in ethanol and passed through the flame several times to sterilize them. Be careful as you do this. An alcohol flame is nearly colorless and therefore invisible. Take precautions to avoid igniting hair or clothing. If you should accidentally set the container of ethanol on fire with the hot instruments, extinguish the flame by replacing the cover on the container.

   Note: Instruments are sterilized by ethanol and not by heat or flame.

7. Rinse the explants in three, 5 minute changes of sterile distilled water. Keep the tissue in its last rinse until you are ready to use it.

8. Transfer the explant from the rinse water to the lid of a sterile petri dish. For leaf tissue, cut the tissue into small squares, not larger than 1 cm^2 using a flamed scalpel. Cut edges that were in direct contact with ethanol and bleach should be trimmed away. Do not use leaf tissue that appears "soaked through" by the disinfectant. Root or stem tissue should be trimmed to remove any tissue that was in direct contact with the disinfect-

Method, continued

ing solutions. Cut the tissue into small cubes, 0.5-1.0 cm^3. In preparing your explants for culture, be mindful of the need to maintain their sterility. Flame your forceps and scalpel frequently and don't lay them down on the work surface after flaming. When working in a laminar flow hood, remember that it is important not to place any contaminated materials up wind of your work. If you are not working in a hood, work quickly and keep your work covered as much as possible.

9. Using flamed forceps, transfer the explants to the culture medium. Be careful to prevent contaminating the medium during this procedure. Explants should be placed firmly in contact with the medium, but should not be buried in it.

10. Label your cultures as to source of explant (type of plant and tissue), date, and your name. Consult the instructor about where cultures are to be incubated.

11. Check the cultures periodically before your next lab class. Any cultures that become grossly contaminated should be autoclaved before disposal.

Results

Examine your cultures and others. Use a microscope. Describe the callus tissue formed and determine the following. How do callus cells compare in appearance to the cells from which they grew? Is there any difference in the appearance of the cultures developed from different plant species or from different tissues of the same plant? Did all of the cultures grow equally well? Did callus tissue form at the same time and grow at the same rate in each of the cultures? Did any of the cultures begin to regenerate roots or shoots? What types of contaminants appeared in the cultures?

Within the first week explants can be expected to swell noticeably. Callus formation should begin during this time. Callus should appear on cut edges of the explant and will look like a rough white fringe of cells. In subsequent weeks, the callus will proliferate into an undifferentiated mass of cells. If desired, the callus can be cultured away from the explant. Remove the clump from the explant tissue with sterile forceps and place it on fresh medium. Whether or not the callus is to be maintained, students should explore the culture with forceps to get a feel for the texture of this tissue. Depending on what plant material is used for the cultures, shoots or roots may begin to regenerate from the callus. This is not a likely outcome, but it is possible.

Results, continued

The major difficulty faced in initiating plant tissue cultures is that of contamination. Plant media are excellent substrates for a host of fungi and bacteria. Generally speaking, bacterial contaminants are more easily dealt with. Antibiotics such as cefotaxime (100 µg/ml) or piperacillin (100 µg/ml) can be included in media for short term bacterial control. If you choose to use antibiotics, remember that they should not be added to media before autoclaving. They should be added as a filter-sterilized stock solution to cooled medium immediately before it is dispensed into tissue culture vessels. Fungal contaminants are nearly impossible to get rid of. For the teaching lab, probably the best advice about contamination is to tolerate it if possible (that is, if the cultures aren't grossly contaminated) over the short term of the experiment and dispose of contaminated cultures by autoclaving. Sometimes, contaminants will not interfere with the outcome of the lab exercise.

UNIT 1: PLANT CELL CULTURE

MODULE 2: PREPARATION AND FUSION OF PROTOPLASTS

Mark Holland

* Introduction
* Safety Guidelines
* Experimental Outline
* Materials
* Pre-lab Preparation
* Method
* Results

Introduction

The objectives of this module are :

1. To prepare a mixed population of protoplasts from red onion.
2. To perform a cell fusion experiment.

Protoplasts are plant cells whose cell walls have been removed. Recall that plant cells in vivo are surrounded by rigid cell walls composed largely of celluloses and cemented together by pectins. The cell wall normally confines the cell, preventing it from bursting as a result of turgor pressure. For some biotechnology and genetic engineering protocols, however, the cell wall is a substantial barrier. Releasing plant cells from the confines of the cell wall allows them to be manipulated by microinjection, electroporation and fusion.

Many techniques for the production of protoplasts have been developed. Originally, tissues were bathed in a solution of high osmotic potential to shrink the cells. Then, cell walls were broken mechanically by cutting or abrasion. Finally, by reducing the osmotic potential of the medium, the cells were made to swell and pop out of the broken cell walls. The yield of viable protoplasts from this type of protocol is generally low, thus, except for some difficult tissues, this strategy is rarely followed. Most methods today depend on the enzymatic breakdown of cell walls first demonstrated by Cocking (1960. Nature 187:927-929). Protoplast digestion mixtures generally include several enzymes, the concerted action of which efficiently releases cells from the cell wall. A mixture of cellulase, pectolyase, and macerase will be used in this experiment. Because naked plant cells are fragile, and can burst in solutions of low osmotic potential, mannitol is included in the digestion mixture as an osmoticum. Other ingredients included improve the stability of the protoplasts.

The plant material used in the experiment is red onion. From this plant, you will prepare protoplasts from the bulb (a modified leaf structure). Because not all cells of red onion are pigmented, protoplasts are a mixed population of colored and colorless. After preparing the protoplasts, you will use them to perform a cell fusion experiment. By treating the protoplasts with solutions of polyethylene glycol and calcium, they can be made to fuse to form hybrid cells. These cells will be recognizable under the microscope. Can you think of any possible applications of this procedure in biotechnology? (Hint: What if the fused protoplasts originated from different genotypes, from different species?)

If your purpose in this experiment were to grow up the hybrid cells or even to regenerate whole plants from them, it would be necessary for you to isolate the protoplasts from sterile (i.e. axenic) tissue, to maintain sterility throughout the protocol, and to return the hybrid

Introduction, continued

cells to culture where they could continue their development. These manipulations are not trivial. In practice, a protoplast fusion experiment of the type you are doing here would be done on a much larger scale and the protoplasts would be treated somewhat differently. First, the protoplasts to be fused would likely originate from different tissues, generally from different genotypes or even different species. The protoplasts would be washed after isolation to remove traces of the digestive enzymes used to remove their cell walls. After fusion, they would be plated in media by techniques that encourage them to regenerate cell walls. In this laboratory experiment, the precautions necessary to maintain your protoplasts in long term culture will not be followed, but you should understand how this experiment would be different if it had different goals.

Safety Guidelines

Follow standard laboratory safety practices.

Experimental Outline

Cut up onion tissue

|

Incubate in digestion mixture (1 hr)

|

Collect protoplasts, observe, allow to settle on slide (20 min)

|

Add fusion buffer and incubate (20 min)

|

Add W10 buffer and observe fusion (20 min)

Materials

Plant material: Red onion. Choose a firm unblemished bulb.

Petri dishes or depression slides: Small diameter (60 mm) disposable dishes or microscope slides with depressions or wells are the container of choice for digesting the plant tissue. Small test tubes will work, but petri dishes offer the advantage that they can be viewed on a microscope so that students can monitor protoplast release.

Microscope: An inverted scope, of the type normally found in tissue culture labs, is the instrument of choice for this use. If such a scope is not available, the normal student scope will work if the experiment is done on a microscope slide. Depending on the magnification available, even a dissecting scope can be used. Since the success of this experiment depends on the students' viewing the results, test available equipment before attempting the lab exercise.

Pasteur pipets

Enzyme solution:

- 1.5% Cellulysin
- 1.5% Macerase
- 0.2% Pectinase
- 0.4 M Mannitol
- 2.0% Glycine

Materials, continued

PEG Fusion Buffer:

0.3 M glucose
66 mM $CaCl_2$
40% polyethylene glycol (molecular weight 1400)-pH 6.0

Solution W10:

9 parts solution A:	0.4 M glucose 66 mM $CaCl_2$ 10% dimethylsulfoxide
1 part solution B:	0.3 M glycine-OH buffer pH 10.5

A kit, which contains all the critical reagents required for this experiment, is available from EDVOTEK (1-800-EDVOTEK).

Pre-lab Preparation

Timetable of Events

Prepare protoplasts (1 hr)
Do fusion experiment (45 min)

Filter the enzyme solution through a 0.45 μm filter (Gelman) before use. The solution can be stored frozen for several months without significant loss of activity. Also prepare some of the same solution without enzymes for use as "medium" in step 6 of the student's experimental procedures.

Method

Adapted from a method of Menczel et al. (1981. Theoretical and Applied Genetics 59:191-195)

1. Using a razor blade or scalpel, cut a small square of tissue (about 1 cm^2) from one of the layers of a red onion. You will notice on the cut edge that the tissue is only red on one surface. Slice the tissue again, layer cake fashion, to expose cells that were buried in the center of the square.

2. Place the half of your tissue containing both red and white cells face down in a small pool (about 0.2-0.5 ml) of enzyme solution. Cover the container in which the tissue is being digested to prevent the enzyme solution from evaporating. Allow the enzyme to work for 1/2 to 1 hour. You will want to examine the digestion periodically during this time to monitor the release of protoplasts. It will help to occasionally agitate the mixture gently.

Method, continued

3. At the end of 1 hour, remove with forceps any remaining bits of undigested tissue. Remove 2 drops of protoplasts to a clean microscope slide. Protoplasts are spherical structures. Do not cover them with a coverslip. Check the slide on the microscope to see that you have both red and colorless protoplasts. The actual number of protoplasts on the slide is not critical, but there should be enough in your sample that they are not difficult to find. Typically, several dozen should be readily visible. If this is not the case, make a new slide. (The protoplasts are actually large enough so that you can see to collect them with a pipet.)

4. Place the slide on the lab bench where it will not be disturbed and allow the protoplasts to settle onto the glass about 15-20 minutes.

5. Add 1 drop of PEG Fusion Buffer to the protoplasts. Check the slide on the microscope at this point. You should see the protoplasts clumping together. Wait 15 minutes, then carefully remove the liquid from the slide with a pipet. The protoplasts should remain behind.

6. Add 2 drops of solution W10 to the slide. Check the slide on the microscope periodically over the next 20 minutes during which fusion may begin to occur. At the end of this time, if no fusion has occurred, add several drops of medium to the slide and view again.

Results

Draw a picture showing: colorless and red protoplasts, fusion products.

Label the following: plasma membrane, vacuole, nucleus.

How can you distinguish between fused and non-fused protoplasts?

Results, continued

If a problem arises, it is likely to be that protoplast yield or yield of fused protoplasts is low. Even under these circumstances, positive results from at least a portion of the class should be forthcoming from the experiment.

Since handling the protoplasts under sterile conditions, as would be done in a research lab, presents difficulties for the teaching lab, the experimental protocol ends with the production of fusion products. Students should understand how the experiment would be performed differently if the goals of the experiment were to maintain the hybrids in culture. This is discussed in the student guide.

The plant material used in the experiment, red onion, was chosen because it is readily available and produces a good yield of protoplasts within the time limitations imposed by a laboratory class. The other benefit of using red onion for this experiment is that digestion of the red onion tissue gives a mixed population of cells, red and colorless. This means that for the fusion experiment included in the module, it is not necessary to prepare two different plant tissues. Other plant materials may perform equally well, but substitutions should be tested before class since all plant tissues do not respond equally to this mixture of enzymes and osmoticum. Some tissues will take from a few hours to overnight to release a sufficient number of protoplasts for the fusion experiment. If desired, protoplasts may be prepared before class and used for the fusion experiment alone during the laboratory period.

UNIT 1: PLANT CELL CULTURE

MODULE 3: REGENERATION AND CLONING

Mark Holland

Introduction

The objective of this module is to test the effect of plant growth regulator on tissue cultured plant cells and to regenerate plants from callus tissue.

In Module 1 of this unit, you investigated the ability of plant cells to de-differentiate into callus tissue in response to wounding. Callus cells placed in an artificial culture medium can be maintained in a persistent undifferentiated state. In this module, you will explore the ability of cultured plant cells to redifferentiate into whole plants. In cell culture, differentiation is effected by plant growth substances (plant hormones) in the culture medium. Cell culture allows us to control the concentration of plant growth regulators and thus affords some control over the differentiation of the plant cells. By changing the hormonal composition of the culture medium we can encourage the growth of roots, shoots, or callus. Two classes of plant growth regulators are used for this purpose, auxins and cytokinins. The specific plant growth regulators used in this experiment are indoleacetic acid (IAA, an auxin) and kinetin (a cytokinin). The media formulations to be tested contain different combinations of these components. As you perform the experiment, be mindful of the correlation between cytokinin/auxin and organ development.

Safety Guidelines

Follow standard laboratory safety practices.

The autoclave or pressure cooker required for media preparation is a burn hazard.

Alcohol or gas flames used to sterilize instruments during the lab class are extremely hazardous. Exercise due care to avoid igniting hair or clothing.

Experimental Outline

1. Prepare plant tissue
2. Make explants
3. Inoculate medium with explants
4. Observe cultures over a period of several weeks

Materials

Plant material

African violet (*Saintpaulia ionantha*) leaves. Choose leaf material that is not damaged, wilted, or bruised.

Culture media

MS (Murashige and Skoog) minimal organics medium to which four different combinations of plant growth regulators have been added:

		IAA (mg/l)	Kinetin (mg/l)
Medium	1	0	0
	2	0.1	10.0
	3	1.0	0
	4	10.0	5.0

You will also need forceps, scalpel, and other equipment for establishing a cell culture as used in Module 1.

Pre-lab Preparation

Timetable of Events

Prepare explants (30 min.)
Inoculate media with tissue (30 min.)
Observe cultures over a period of several weeks

This module is an extension of the techniques and principles presented in "Establishing a Plant Cell Culture" (Module 1). In this experiment, cell cultures will be established on four different media, each containing a different combination of the plant growth regulators indoleacetic acid (IAA, an auxin) and kinetin (a cytokinin). In general, high cytokinin/auxin favors the formation of shoots, while low cytokinin/auxin favors root growth. While endogenous hormone levels in tissue explants can introduce some variability into the experiment, the protocol is reliable and produces dramatic results for the students. As an extension of Establishing a Plant Cell Culture activities, callus produced in that laboratory exercise can be used as an explant in this exercise.

Refer to "Establishing a Plant Cell Culture" for background and details on plant culture technique.

Plant material

African violet (*Saintpaulia ionantha*). The plant is easy to obtain and familiar to the students. Choose a healthy plant with firm, unblemished leaves. If plants with variegated leaves are used, regenerated shoots will be of two types, green and albino. This oddity emphasizes that regenerants come from single cells and that the green and white sectors of the leaf have different plastid genotypes; the plant is a genetic chimaera. Of course albinos can only be maintained in culture, but sometimes they can be grown to quite a large size. Other plant material can be used for this experiment. The best of alternatives are probably tobacco (use leaf as explant) and cauliflower (use small florets as explant).

Media:

Prepare MS minimal organics medium (with added sucrose and agar). Divide the medium into four aliquots. Add growth regulators from stock solutions (also described in Module 1) according to the following schedule:

Pre-lab Preparations, continued

		IAA (mg/l)	Kinetin (mg/l)
Aliquot	1	0	0
	2	0.1	10.0
	3	1.0	0
	4	10.0	5.0

After adding the plant growth regulators, dispense the media into flasks or tubes for culture and autoclave. Note: If equipment/ supplies for media preparation are limiting, an experiment similar to this one is available ready-made from Carolina Biological Supply Co.

Method

1. Prepare leaf tissue from African violet for cell culture as practiced in module 1. Briefly, disinfect leaf tissue by sequential washing, soaking in 70% ethanol, soaking in 10% household bleach, and rinsing in three changes of sterile distilled water. Refer to methods in "Plant Cell Culture I" for complete directions if necessary.

2. Maintaining the sterility of the leaves, cut explants (about 1 cm^2) from the leaf blade or about 0.5 cm long from the petiole.

3. Distribute the explants among the four different media to be tested. Label the cultures with your name and date (include type of medium if this information is not already on the culture). Incubate your cultures under fluorescent lights (16 hr day, 26°C or room temperature).

4. Check the cultures at least once per week, observing the development of the cultures and looking for contamination by bacteria or fungi. Grossly contaminated cultures should be autoclaved before disposal. A small amount of contamination might be tolerated if it doesn't take over the culture.

5. The cultures should be allowed to develop for 4-6 weeks.

Results

Which of the cultures produced callus, roots, shoots? What cytokinin/auxin (kinetin/IAA) ratios favor shoot and root formation?

To continue the experiment, shoots can be excised and transferred to a medium with no hormones. This should encourage root growth. Rooted plants can eventually be transferred into soil if high humidity is maintained around the plant while it is becoming established. Cover the plant with a plastic bag or inverted jar to accomplish this. Explants that do not form shoots can be transferred to a medium that encourages shoot formation.

UNIT 2

Genetic Transformation in Plants

UNIT 2: GENETIC TRANSFORMATION IN PLANTS

MODULE 4: USE OF *AGROBACTERIUM TUMEFACIENS* AS A CLONING VECTOR

Mark Holland

* Introduction
* Safety Guidelines
* Experimental Outline
* Materials
* Pre-lab Preparation
* Method
* Results

Introduction

This is the first of two modules related to genetic transformation in plants. The objective of this first module is to produce a crown gall tumor by infecting a plant with wild type *Agrobacterium tumefaciens*, and then to assay opine production in the tumor tissue.

In the next module, the objective is to infect plant tissue (a radish slice) with an engineered strain of *Agrobacterium* and assay the activity of a marker gene (GUS) in the transformed tissue.

Agrobacterium tumefaciens is a plant pathogen that causes crown gall disease in dicotyledonous plants. The disease is characterized by a tumorous growth, generally at the soil line, on the stem of the infected plant. Molecular studies of the disease reveal that crown gall tumors develop as a result of a genetic transformation of the plant by the bacterium. That is, during the infection process, the bacterium makes a heritable change in the DNA of the plant and that change results in the formation of a tumor by the plant.

Crown gall disease is of interest to plant molecular biologists and to biotechnologists because it is an example of naturally-occurring genetic engineering and has led to the development of useful tools for directing genetic change in plants. The laboratory exercises included in these two modules are designed to illustrate:

1. the development of crown gall disease following infection by *Agrobacterium* and demonstration of altered metabolism in tumor tissue, and

2. an example of how the *Agrobacterium* system has been adapted by molecular biologists to allow them to produce genetically transformed plants in the laboratory.

Highlights of the Biology of Crown Gall Tumor Formation

1. *Agrobacterium tumefaciens* is a soil bacterium and an opportunistic pathogen of plants. In response to substances released from a wound on the plant, genes in the bacterium that condition the infection process (the so-called virulence genes) are induced. Acetosyringone, which is included in the bacterial culture medium for the second experiment, is a phenolic compound from plant cell wall that induces the virulence genes of *Agrobacterium*. Including this compound in the culture medium improves the rate of infection of plant cells by the bacterium.

Introduction, continued

2. In addition to its genomic DNA, *Agrobacterium* carries a large plasmid. The plasmid is called "Ti" or "tumor-inducing" plasmid. During infection, the bacterium transfers this plasmid to the plant cell.

3. A portion of the Ti plasmid, the T-DNA, is then inserted into the plant DNA by genetic recombination. The plant is genetically transformed. The ends, or "borders" of T-DNA are functionally analogous but not structurally homologous to the transposons of bacteria and plants. All of the details surrounding insertion of T-DNA at its borders into the plant genome are not known.

4. Genes within the T-DNA are transcribed and translated by the plant to produce tumorous growth. What are these genes? First, there are genes for the production of cytokinins. These are plant hormones affecting cell division. Transformation of the plant with cytokinin biosynthetic genes produces the unregulated cell division that results in the development of a tumor. (For additional information about plant hormones and the regulation of cell division, refer to the unit on Plant Cell Culture elsewhere in this series.) Second are genes for the synthesis of non-protein amino acids called opines, which can be used as a source of carbon by the bacteria and which probably also influence bacterial mating. The result of the genetic transformation of the plant by the pathogen is that the plant is forced to produce a novel resource, opines, for the bacterium.

Safety Guidelines

Agrobacterium tumefaciens is a plant pathogen and a serious pest on some crop plants. All cultures, and contaminated materials (including syringes, blades, swabs, etc.) should be autoclaved before disposal.

Experimental Outline

Experiment 1 (Week 1)

Inoculate plant with bacterium (15 min)
Observe tumor formation over a period of several weeks

Plant material - *Kalanchoe* or alternative
Syringe or razor blade/scalpel
Cotton swabs
Labels (tie-on type tags)
Scotch tape
Culture of wild type *A. tumefaciens*

Using a syringe needle or razor blade, make shallow wounds in the surface of the plant leaf, petiole, and/or stem. Using syringe or cotton swabs, inoculate the wound sites with the bacterial culture. Cover the inoculation site with a piece of clear tape. Attach a label bearing the date and your name to the inoculated leaf. Return the plant to the growth chamber, greenhouse, or window sill and observe tumor formation over a period of two to three weeks.

Materials

Plant tumor tissue
Razor blade or scalpel
Microcentrifuge tube (one for each tumor assayed)
Glass rod (3 mm diameter)
Sterile sand
Apparatus for paper chromatography
Hair dryer
Ninhydrin spray
Microcentrifuge

Pre-lab Preparation

Timetable of Events

Experiment 1 requires a short period of class time for inoculation of the plants with the bacterium. Tumor formation should be expected to take from two-three weeks. Analysis of opines will generally occupy an entire lab period, although during wait time (for example, while paper chromatogram develops) other activities could be planned.

Experiment 1 - Inoculate 50 ml YEP medium with *Agrobacterium tumefaciens* (wild type) and incubate on shaker overnight at 26-28°C. (Note that this temperature is not what many people are used to for bacterial cultures.) Obtain plant material.

Equipment and Facilities

Experiment 1

- Autoclave
- Shaker at 26-28°C
- Apparatus for Paper Chromatography
- Hair dryer

Materials/Supplies

Experiment 1

- Plant material - *Kalanchoe*
- *Agrobacterium tumefaciens* wild type culture
- Medium for bacterium (YEP)
  - YEP Medium (per liter)
    - 10g Bacto Peptone
    - 10g Bacto Yeast Extract
    - 5 g NaCl
- Syringes or razor blades/scalpels
- Cotton swabs
- Scotch tape
- Labels
- Microcentrifuge tubes (1.5 ml)
- 3 mm diameter glass rods
- Sterile sand
- Chromatography paper
- Ninhydrin spray

Method

Remove a small piece of each tumor to be assayed from the plant (not more than 0.1 g) and transfer each to its own microcentrifuge tube. Also include a sample of plant tissue that is not part of a tumor. This is a control. For each sample, moisten the tip of a glass rod with water, touch on sand to pick up a small amount, and use to grind tumor or control tissue in its tube. Spin the tubes in the microcentrifuge for five minutes. Spot the supernatant on a paper chromatogram and air dry. Include a standard of nopaline (2-5 μl of a 500 μg/ml stock). Develop the chromatogram in a chromatography tank of 4 : 1 : 1 butanol : acetic acid : water. Remove the chromatogram from the tank and air dry (a hair dryer will speed this up). Stain the chromatogram with a spray of ninhydrin to reveal whether your tumor tissue is producing nopaline.

Results

Describe tumor formation in your plant tissue. Use diagrams if they will help. What was the result of your opine assay? Were all tissues analyzed apparently transformed? Was opine production variable in the plant?

UNIT 2: GENETIC TRANSFORMATION IN PLANTS

MODULE 5: ASSAY OF GUS GENE IN TRANSFORMED PLANT TISSUE

Mark Holland

* Introduction
* Safety Guidelines
* Experimental Outline
* Materials
* Pre-lab Preparation
* Method
* Results

Introduction

This module is a continuation of the previous module entitled "Use of *Agrobacterium tumefaciens* as a Cloning Vector." Because introductory information was presented at the beginning of Module 1, the reader should refer to the information provided in the previous section for review. Not all information presented in Module 1 will be repeated with this continuation module.

Taming *Agrobacterium* for Use in Biotechnology

Molecular biologists have "domesticated" *Agrobacterium* to make it useful in the laboratory as a tool for plant transformation. To appreciate how the bacterium and its plasmid have been engineered, consider the example of the *Agrobacterium* strain used in experiment 2 below.

Agrobacterium tumefaciens C58 is a nopaline-producing strain. In engineering this strain for use as a tool for plant transformation, the Ti plasmid has been replaced with 2 smaller plasmids. This type of construction is referred to as a "binary vector". One of the binary vector plasmids (here called pTiC58) is a shortened version of the Ti plasmid. From it have been removed the T-DNA and genes for cytokinin biosynthesis. C58 containing pTiC58 is thus said to be disarmed. That is, it is incapable of producing tumorous growth. The second of the binary plasmids (here, pMSG) contains T-DNA which has been altered by the insertion of a marker gene, a gene whose presence and expression in plant tissue is taken as evidence of transformation. The marker gene in pMSG is the GUS, or beta-glucuronidase gene. GUS has several properties that make it useful as a marker. First, the enzyme beta-glucuronidase is not normally found in plants. This means that false positive results are a rare event. Second, different substrates are available for the enzyme that produce either a visible histochemical stain (used in this module) or a fluorescent product detectable at lower levels upon cleavage by the enzyme. In addition to the features already mentioned, each of the plasmids in the binary system carries genes for resistance to antibiotics. pTiC58 makes the bacterium resistant to spectinomycin, while pMSG conditions resistance to tetracycline. The inclusion of these two antibiotics in the bacterial culture medium insures that only bacterial cells containing both plasmids will grow. During the transformation process, only the T-DNA and the GUS gene cloned into it are transferred to the plant cell and integrated into the plant genome.

Safety Guidelines

Agrobacterium tumefaciens is a plant pathogen and a serious pest on some crop plants. All cultures and contaminated materials (including syringes, blades, swabs, etc.) should be autoclaved before disposal.

Experimental Outline

Experiment 2

Inoculate radish slices with *Agrobacterium*
Experiment 2, Day 2
Stain radish slices for GUS activity

Materials

Experiment 2 (from L.A. Castle and R.O. Morris. 1990. Plant Molecular Biology Reporter 8(1):28-39)

Materials:

- *A. tumefaciens* culture (strain C58- contains the plasmids pC58 and pMSG)
- Radish roots
- Razor blade or scalpel
- Petri dish of plant medium
- X-gluc stain

Note: This strain of *Agrobacterium* can be obtained by contacting Dr. Mark Holland, Dept. of Biology, Salisbury State University, Salisbury, MD 21801. Before requesting the bacteria, you must contact your State Department of Agriculture for information on regulations that apply to distribution and use of this plant pathogen.

A kit, which contains all the critical reagents required for this experiment, is available from EDVOTEK (1-800-EDVOTEK).

Materials, continued

Equipment - Experiment 2

- Autoclave
- Spectrophotometer
- Shaker as above
- Incubator at 37°C
- Dissecting Microscope

Supplies - Experiment 2

- Plant material - Radish root
- *Agrobacterium tumefaciens* strain
- Bacterial media
  - YEP (recipe above)
  - M9+ (per liter)
  - 200 ml 5x Salts (recipe below)
  - 50 ml 20% glucose
  - 12.5 ml 20% casamino acids
  - 1 ml 1M $MgSO_4 \cdot 7H_2O$
  - 0.1 ml 0.5% Thiamine HCl
  - Filter-sterilize glucose and thiamine through a 0.45 µm filter. Autoclave the rest separately. Mix the components, adjust pH to 5.5 and filter-sterilize.

  5x Salts (per 200 ml)

  - 2.7 g Na_2HPO_4
  - 15 g KH_2PO_4
  - 25 g NaH_2PO_4
  - 5 g NH_4Cl
  - 2.5 g NaCl

X-gluc stain (50 ml) Make fresh before use:

- 25 ml 0.2M NaH_2PO_4 buffer, pH 7.0
- 23.5 ml sterile water
- 0.25 ml 0.1M $K_3Fe(CN)_6$
- 0.25 ml 0.1M $K_4Fe(CN)_6 \cdot 3H_20$
- 1 ml 0.5M Na_2EDTA
- 50 µL x-gluc stock solution *
- Store at 4°C in the dark.

*Make a stock solution of 0.1 mg/µl x-gluc in N,N-dimethylformamide.

Other Solutions

- 0.5 M Acetosyringone in DMSO
- antibiotics for plasmid selection
- 70% ethanol
- 5% bleach
- Sterile water
- MS salts
- 0.75% agar, 0.5x MS plates

Pre-lab Preparation

Timetable of Events

Experiment 2 requires a short lab period for preparation of tissue and inoculation of the tissue. Students need to return to the lab briefly after the lab to place tissue in stain and to observe results. Both experiments can be done during the same laboratory period. Preparation of radish slices and inoculation of them with the bacterium can be done during opine analysis lab.

Experiment 2

DAY ONE

Grow 50 ml overnight culture of *Agrobacterium* (engineered strain) in YEP medium at 26-28°C.

DAY TWO

Inoculate 5% into fresh medium. Include antibiotics (spectinomycin 50 µg/ml and tetracycline 50 µg/ml) in the medium for plasmid selection. Grow 4-6 hours (mid-log phase). Use this culture to inoculate 2% into M9+ medium. Add acetosyringone to a final concentration of 200 µM. Usually 100 ml of culture will be required to incubate 10-15 radish slices.

DAY THREE

After 24 hours of growth in M9+ medium, check the density of the culture in the spectrophotometer (A_{600}). Spin the cells down at 600xg for 10 minutes at 15°C. Gently resuspend in one-fourth of the original volume of M9+ (no antibiotics, 50 µM acetosyringone) and spin again as previously. Resuspend in M9+ to an A_{600} of 1.2. Prepare solutions for GUS staining and pour plates for incubating radish slices.

Method

Surface-sterilize the radish root in 70% ethanol for one minute followed by 5% bleach for five minutes. Rinse in sterile water for five to ten minutes, changing the water several times. Slice the radish cross-sectionally into disks one to two millimeters thick using a sterile blade (scalpel or razor blade can be sterilized by dipping into 95% ethanol and passing through a flame). Place the disks in a sterile petri dish and pour a suspension of the bacterial culture over the disks to cover them. Cover the dish and let stand for 30 minutes. Rinse each disk in sterile bacterial medium, blot lightly on sterile paper towels and place them in the dish of plant medium. Seal the dish with a piece of parafilm and store right side up in the dark for 48-72 hours.

After 48-72 hours: Remove the radish disks from the plant medium, rinse them with 0.1 M phosphate buffer and place them into a clean petri dish. Fill the dish with enough X-gluc stain to cover the tissue. Cover the dish, seal with parafilm to retard evaporation. Incubate the dish at 37°C for 1-4 hours (or overnight) while color develops. Observe the pattern of expression of the marker gene as evidenced by the appearance of an intense indigo blue color. You may want to look at the tissue under a dissecting microscope.

Results

What tissues of the radish root expressed GUS activity?

The expression of the GUS gene shows some tissue specificity. Activity is detected only in and around vascular tissues. Thus, a single band of tissue just beneath the red skin of the radish root, is expected to stain deep blue.

UNIT 3

Cell Fractionation in Plants

UNIT 3: CELL FRACTIONATION IN PLANTS

MODULE 6: DENSITY-GRADIENT CENTRIFUGATION

Richard Echols
Edward Kisailus

* Introduction
* Safety Guidelines
* Materials
* Pre-lab Preparation
* Method
* Results

Introduction

The objective of this module is to demonstrate high speed density gradient centrifugation with low speed centrifuge in a model system. Centrifugation is a simple operation used in almost all biological separations. It fractionates by applying a centrifugal field to the mixture to be separated instead of gravity. Strategies for cell fractionation are based on physical properties of organelles.

Because organelles differ from one another in mass and density, they are efficiently separated from one another by centrifugation.

If any two particles are suspended in water, you might imagine that the larger or more dense of the two would fall (sediment) through the solution more quickly than the other in response to gravity (a downward force) and resistance provided by the water (an upward force). When the particles in question are the size of organelles, however, this sedimentation takes a very long time. The basic idea behind centrifugation is that in the centrifuge, forces (centrifugal forces) are applied to particles in solution that are much greater than the force of gravity. In a centrifugal field, therefore, sedimentation occurs more quickly than in a gravitational field. Because the force of a centrifugal field on particles suspended in the centrifuge tube replaces gravity as the principle force acting on them, the strength of a centrifugal field is reported as "times the force of gravity" or "x g". Two centrifugation techniques are central to most cell fractionation protocols. These are differential centrifugation, and isopycnic density gradient centrifugation.

In differential centrifugation, a cell homogenate is centrifuged several times, each time at a greater speed and for longer periods of time. In early spins, only the largest particles sediment (or pellet) to the bottom of the centrifuge tube. Such particles might be whole cells, fragments of cell wall, nuclei, or grains of starch. Later, chloroplasts and mitochondria pellet. At the highest speeds, and these might be tens of thousands of revolutions per minute, microsomes and ribosomes, even dissolved protein molecules, can be pelleted.

For isopycnic density gradient centrifugation, aqueous solutions of different densities are prepared by dissolving sucrose or other solutes in them. Before centrifugation, these solutions are layered in the centrifuge tube from most to least dense (bottom to top). Gradients of this sort can be prepared in discrete layers (so-called step gradients) or as a single layer of continuously changing density. The cell homogenate to be fractionated is then layered on top of the gradient and centrifugation is carried out. During the spin, particles suspended in the homogenate sediment. Rather than pellet, however, they stop moving through the gradient as they enter a layer whose density matches their own.

Introduction, continued

Other centrifugation techniques have been developed. The choice among them is governed by many factors, not the least of which are the practical, "What works best for my experiment?" and "What equipment do I have available in the laboratory?"

In density gradient centrifugation, the particles have to reach sedimentation equilibrium. The technique can yield data about molecular weight, density, shape and purity of a given group of molecules. The operation is simple in principle, but more complex in practice. There are several materials that can serve as density gradient supports, for the preparation of gradients in the density range of 1.00 g/ml to 1.4 g/ml. These include cesium chloride, sucrose (sugar), ficoll and percoll. The density gradient is prepared in the centrifuge tube, and the sample is layered and centrifuged. The components (macromolecules) migrate until they reach the zone or equilibrium position, where their density equals the density of the medium.

Density gradients can be constructed in centrifuge tubes as discontinuous or continuous gradients. A discontinuous gradient is constructed by loading a centrifuge tube with layers of varying densities of the separating medium. This experiment will use percoll as the density gradient medium. The sample to be separated is added to the top layer of the medium and centrifuged at high speed. The components of the sample will sediment and form a band at the interface of the layer which matches their density. A continuous gradient is formed when a medium with a uniform density is exposed to a centrifugal force.

The separation medium is a suspension of small particles which, when placed under a high centrifugal field, concentrate toward the bottom of the centrifuge tube forming a gradient. As with the discontinuous gradient, the cells or subcellular particles to be separated are layered above the separating medium, sucrose or percoll, and centrifuged at high speed. The particles will sediment at their respective densities in the self-generated density gradient.

Percoll is a colloid (finely divided particles suspended in an uniform medium) of silica particles 15-30 nanometers in diameter which have been coated with polyvinyl pyrrolidone (PVP), thus increasing their stability as a colloid and eliminating their toxicity to cells. Sucrose can be used in this model experiment, but cannot be used to separate cells and some cell organelles. In this experiment, density gradient markers are substituted for biological particles.

Safety Guidelines

Follow standard laboratory safety procedures.

Materials

15 ml centrifuge tubes

Percoll solution of different densities

Centrifuge (capable of 1000 rpm)

Density gradient marker beads

A kit, which contains all the critical reagents required for this experiment, is available from EDVOTEK (1-800-EDVOTEK).

Pre-lab Preparation

Notes to Instructor:

1. In calculating the sedimentation number x g, the angle and radius of the rotor must be taken in consideration.
2. In differential centrifugation impure materials are isolated because the small particle in the lower part of the tube is spun down before the large ones at the top of the tube
3. Sucrose cannot be used for cells and some cellular organelles because it alters osmotic pressure.
4. The discontinuous gradient will gradually become continuous on standing. This effect will account for difference in the R_f value for solutions of beads allowed to reach equilibrium over night vs those centrifuged in 25 min.
5. There should be no differences in R_f value if the solution is centrifuge for 50 min. instead of 25 min.
6. In a pure water solution all the beads should settle to the bottom of the tube.
7. If a density of 1.10 solution is used all of the beads should remain on the surface of the medium.
8. Typical densities of subcellular particles in sucrose are: nuclei 1.32, mitochondria 1.19, chloroplasts 1.22 golgi membranes 1.12, lysosomes 1.21 and ribosomes 1.60.
9. Density beads may be washed and re-used.

Pre-lab Preparation, continued

Preparation of the Density Gradient

1. Formula used to prepare the desired densities of percoll
   Vy = Vi {(Di - D)/(D - Dv)}

   Vy = volume of diluting media (RPMI 1640 with L-glutamine) or Normal saline media

   Vi = volume of stock Percoll

   Di = density of stock solution (percoll density = 1.130 g/ml)
   Dv = density of diluting media (density =1.002 g/ml)
   D = density of diluted solution produced

2. Prepare a working Percoll solution at a density of 1.100 g/ml.
   Vy = Vi {Di-D)/D-Dv)}

   = 50 ml {(1.130 g/ml - 1.100 g/ml)/1.100 g/ml-1.002 g/ml)}
   = 50 ml of stock percoll added to 14.79 ml of media will produce 64.79 ml of working percoll at a density of 1.100 g/ml

3. Other concentrations will be diluted from the working percoll (i.e. using the working percoll as the stock percoll in the above formula).

4. Other densities:

   a) 1.05 g/ml 5 ml working percoll + 5.21 ml media + 1 drop yellow food coloring)

   b) 1.06 g/ml 5 ml working percoll + 3.45 ml media + 1 drop water)

   c) 1.065 g/ml 5 ml working percoll + 2.7 ml media + 1 drop green food coloring)

   d) 1.075 g/ml 5 ml working percoll + 1.7 ml media + 1 drop red food coloring)

   e) 1.08 g/ml 5 ml working percoll + 1.3 ml media + 1 drop blue food coloring)

5. Refrigerate the density gradient solutions after preparation.

Method

Layer different densities of the media into a 15 ml centrifuge tube (the media will prepared by the assistant). Place the most dense layer on the bottom using a 1 ml pipet. Add the layers on top of each other very slowly, being careful not to mix the different densities. You will be able to see the interface on the layers (one drop of food coloring has been added to each solution to facilitate visualization of the various layers). The densities to be used will depend on the density of the marker beads.

1. The percoll densities of 1.080, 1.075, 1.065, 1.060 and 1.050 g/ml are layered in the centrifuge tube in a ratio of 1:1:1:3:1 ml. The appropriate marker beads for this gradient are: 3 green (1.048 g/ml), 4 red (1.062 g/ml), 5 blue (1.072 g/ml) and 6 orange (1.088 g/ml).

2. Add approximately 10 microliters of different density marker beads to the top layer in the tube.

3. To a second centrifuge tube add the first three of the density layers, then add the marker beads, and then add the other density layers.

4. Centrifuge both tubes at approximately 1000 rpms for 25 minutes.

Results

1. Record your observations.

2. Calculate R_f value for each band of beads. Measure in cm the total length of the solution in the tube, this is your "y" value. Next measure in cm the length from the bottom of the tube up to each of the bands. These are the "x" values. Dividing X/Y yields the R_f value for each band.

UNIT 3: CELL FRACTIONATION IN PLANTS

MODULE 7: ISOLATION OF CHLOROPLASTS

Mark Holland

* Introduction
* Safety Guidelines
* Experimental Outline
* Materials
* Pre-lab Preparation
* Method
* Results

Introduction

In the experiments described in the next four modules, differential centrifugation and density gradient centrifugation will be used to isolate two organelles, chloroplasts and mitochondria, from pea seedlings.

Objectives:

1. Isolate chloroplasts from pea.
2. Assay chlorophyll content of isolated chloroplasts.

Chloroplasts are the subcellular sites of photosynthesis, the process by which green plants, using energy from light, produce carbohydrate and oxygen from carbon dioxide and water. Under the microscope, chloroplasts are recognized as bean-shaped, membrane-bound, green (chlorophyll-containing) organelles. In this experiment, chloroplasts will be isolated from a cell homogenate by density gradient centrifugation. A measurement of the amount of chlorophyll in the preparation will be used to assess yield of intact chloroplasts. Finally, the chloroplasts will be saved for the isolation of DNA.

Safety Guidelines

Follow standard laboratory safety practices.

Experimental Outline

Prepare sucrose gradients (30 min)

Harvest pea tissue and process sample (30 min)

Apply sample to gradient and centrifuge (30 min)

Retrieve chloroplasts from gradient and assay yield (30 min)

Materials

Pea seedlings
Homogenization buffer
- 10 mM KCl
- 1mM $MgCl_2$
- 1% (w/v) Dextran T40
- 1% (w/v) Ficoll
- 0.1% (w/v) Bovine Serum Albumin
- Make to volume with 30% (w/w) sucrose in 0.1M Tricine buffer, pH 7.5

Ice bucket
Cheesecloth and Miracloth (Calbiochem)
Centrifuge tubes
Sucrose (gradient) solutions (w/w, prepared in 0.1M Tricine buffer, pH 7.5) 60%, 50%, 40%, 30%
Pasteur pipettes
Blender
Centrifuges (clinical and super speed)
Spectrophotometer

A kit, which contains all the critical reagents required for this experiment, is available from EDVOTEK (1-800-EDVOTEK).

Pre-lab Preparation

Timetable of Events

The exercise can be divided into four parts: 1) preparation of sucrose gradients, 2) sample preparation, 3) centrifugation, and 4) analysis of results. Each of these activities will require approximately 30 minutes.

Equipment Requirements

The protocol described here is written to be used with a refrigerated, super-speed centrifuge (e.g. Sorvall RC series) equipped with a swinging bucket rotor (e.g. Sorvall HB4).

Other comparable centrifuges will perform equally well, but specified speeds and times of centrifugation might have to be changed slightly. The operations manuals for your particular centrifuge and rotor will explain these changes.

The assay of chlorophyll content in gradient fractions requires the use of a spectrophotometer or colorimeter.

A blender is required for tissue homogenization. A standard household blender will work well, but the blades should be as sharp as possible (some laboratories have been known to modify their blenders by replacing the standard blades with razor blades!).

Plant Material: The protocols in this unit all call for the use of pea seedlings as starting material. The seed is relatively inexpensive and easy to grow. For best results, use fresh seed. Seedlings should be about seven days old for the lab. To plant, soak seed overnight in a large container. Sow the seeds on a layer of about 1.5 inches of wet horticultural grade vermiculite in a standard nursery flat (21 x 10 x 2 inches). The seed can be sown thickly - nearly touching one another. (We usually use a 500 ml beaker-full of dry seed to plant one flat. At harvest, you can expect about 300 g of shoot tissue from such a planting.) Cover the seed with 0.5-1 inch of vermiculite and water well. Cover the flat with plastic wrap to hold in moisture until the seedlings begin to emerge. Once the seedlings have emerged and the plastic has been removed, keep well-watered. The seedlings can be grown in the lab on a window sill, in a growth chamber or in the greenhouse. For best results, however, do not grow the seedlings under intense light. Under very bright lights, chloroplasts tend to accumulate large granules of starch and these can do damage during blending and centrifugation. (Note: Left over shoots can be air dried or freeze dried and used for isolation of genomic DNA; see Module 8.)

Pre-lab Preparation, continued

Solutions Required

Homogenization Buffer
10 mM KCl
1 mM $MgCl_2$
1% (w/v) Dextran T40
1% (w/v) Ficoll
0.1% (w/v) Bovine Serum Albumin
Make to volume with 30% (w/w) sucrose in 0.1M Tricine buffer, pH 7.5

Sucrose (gradient) Solutions
60%, 50%, 40%, and 30% sucrose (w/w) in 0.1M Tricine buffer, pH 7.5

Method

The method is adapted from method of: B.J. Miflin and H. Beevers, 1974. Plant Physiol. 53: 870-874.

All solutions should be ice-cold. Keep solutions, samples, gradients, etc. on ice while you are working.

1. Prepare two sucrose gradients in 50 ml centrifuge tubes.

   A. Pipette 5 ml 60% sucrose solution into bottom of each tube.

   B. Layer 5 ml 50% sucrose, then 10 ml 40% sucrose into each tube. Layers should be distinct from one another if you are careful. (Hint: Tip the tube as you add each layer of sucrose. Let tip of the pipette just touch the surface of liquid in tube.)

   C. With the tip of a pasteur pipette or stirring rod, gently mix at the interface of the 50% and 40% layers to diffuse slightly.

   D. Layer 5 ml 30% sucrose on top of the gradient.

   E. Keep the gradients on ice while you prepare tissue sample.

2. Harvest 5 grams of 7-day-old pea seedlings at the soil line with a razor blade.

3. Chop the tissue into small pieces with a razor blade or scissors and transfer them to a chilled blender containing 20 ml ice-cold homogenization buffer. (Note: More than one 5 g batch of seedlings can be blended at a time. If several lab groups are sharing the blender, they should use 20 ml of buffer for each 5 g of seedlings homogenized, then divide the homogenate.)

Method, continued

4. Homogenize with five 2-3 second bursts of the blender at high speed.

5. Filter the homogenate into a beaker (on ice) through four layers of cheesecloth, squeezing the cloth gently to remove most of the liquid (wear gloves).

6. Re-filter the first filtrate through one layer of Miracloth, moistened in homogenization buffer, by gravity. Do not squeeze. You may want to prepare a wet-mount slide of the residue left in the cheesecloth or Miracloth for the microscope. What have you removed from the homogenate by filtration?

7. Layer 10 ml of the filtrate onto the top of each of your gradients (prepared in step 1). Check to see that the two gradients are balanced against one another. If necessary, add homogenization buffer to make the tubes balance. Centrifuge at 4°C in HB4 rotor, 4000 rpm for 5 minutes, then increase speed to 10,000 rpm for 10 minutes. Allow the centrifuge to coast to a stop. Carefully remove your gradients from the rotor.

8. You should see two green bands in the gradient. The green band toward the bottom of the tube is the fraction containing intact chloroplasts. Remove the top of the gradient carefully with a pasteur pipet. Save the two chlorophyll-containing fractions in clean tubes on ice.

9. Prepare wet-mount slides of the chlorophyll-containing fractions and examine on the microscope. What differences do you notice between them?

10. Assay chlorophyll content of the two green bands. For each sample:

    A. Into a clean centrifuge tube, pipette 50 µL of the gradient fraction to be assayed and 0.95 ml distilled water.

    B. Add 4 ml acetone.

    C. Centrifuge in a clinical centrifuge, 5 minutes.

    D. Measure the absorbance of the solution in a spectrophotometer at 652 nm. (The appropriate blank for this measurement is 80% acetone in water.)

    E. Calculate chlorophyll content:
    $A_{652} \times 29$ = µg chlorophyll/10 µl chloroplast fraction.

11. Freeze the intact chloroplast fraction to save for DNA isolation.

Results

Make a diagram of your gradient. Have materials other than chloroplasts banded in the tube? Where are they? What do they look like? Why was the cell homogenate filtered before being loaded on the gradient? What was removed by filtration? What fraction of the chloroplasts in the homogenate have you isolated intact?

After centrifugation, the gradients should show two bands of chlorophyll. The upper of the two contains broken chloroplasts and membrane fragments. The lower of the bands (about three quarters of the distance to the bottom of the tube) contains intact chloroplasts. Probably the most frequent cause of a poor yield of intact chloroplasts is over-blending. Using any type of homogenizer to disrupt tissue is a trade-off between efficiency in breaking cells open and generating so much shear in the solution that organelles are also disrupted.

If you plan to use the chloroplasts from this prep for DNA isolation (Module 9), avoid contaminating the intact chloroplast fraction with broken chloroplasts. Also, try to retrieve the intact chloroplast fraction from the gradient in as small a volume as possible.

UNIT 3: CELL FRACTIONATION IN PLANTS

MODULE 8: ISOLATION OF GENOMIC DNA

Mark Holland

* Introduction
* Safety Guidelines
* Experimental Outline
* Materials
* Pre-lab Preparation
* Method
* Results

Introduction

The objectives of this module are:

1. To isolate DNA from shoot tissue of Pea (*Pisum sativum*).

2. Compare it to DNA isolated from chloroplasts of Pea.

Isolation of DNA from plant tissues is at the heart of plant molecular biology. Because plant cells are surrounded by rigid cell walls and because plant tissues often contain a variety of secondary metabolites that can damage DNA. Thus, DNA isolation from plants can present some particular difficulties. Among the many protocols developed for DNA isolation from plants, the method presented here is one of the simplest and most effective for a variety of plant species.

DNA isolated by methods such as the one presented here represents total cellular DNA. In plants, this means the isolation of three distinct genomes:

1) the nuclear genome
2) the chloroplast genome
3) the mitochondrial genome.

Generally, when we talk about plant DNA we mean nuclear DNA, but it is important to remember that chloroplasts and mitochondria each have distinct genomes. These genomes, like the nuclear genome, are targets of research in molecular biology and genetic engineering. Total plant DNA isolated here will be compared with DNA isolated from chloroplasts in another laboratory exercise.

Safety Guidelines

Reagents used in this protocol are potentially dangerous. CTAB, cetyltrimethylammonium bromide (in the extraction buffer) is a strong detergent and can cause burns to the skin. Chloroform is toxic by inhalation or on contact with skin. Follow your instructor's direction for proper handling and disposal of these materials.

Experimental Outline

Day 1. DNA isolation (2 hrs)
Grind tissue and suspend in extraction buffer (EB)
Incubate at 65°C for 1 hr
Extract with chloroform
Precipitate DNA with isopropyl alcohol
Resuspend the DNA in buffer and store

Day 2. Compare this DNA preparation with DNA isolated from chloroplasts by gel electrophoresis. (Module 9)

Materials

EB (extraction buffer): 50 mM Tris pH 8.0, 1% CTAB, 50 mM EDTA, 1 mM 1,10-O-phenanthroline, 0.7 M NaCl, 1% beta-mercaptoethanol

Chloroform

Isopropyl alcohol

80% ethanol, 15 mM ammonium acetate, pH7.5

TE buffer: 10 mM Tris pH 8.0, 1 mM EDTA

Centrifuge Tubes (50 ml capacity, capped)

Water bath 65°C

Centrifuge

A kit, which contains all the critical reagents required for this experiment, is available from EDVOTEK (1-800-EDVOTEK).

Pre-lab Preparation

Pea tissue used in this protocol should be air dried or freeze dried. Grow the plants according to directions in Module 2. Tissue can be air dried in the laboratory by spreading the cut shoots on a sheet of absorbent paper and turning daily. Tissue treated in this way will typically dry in 5-7 days. Dried tissue can be saved for use at a later date by sealing in a tightly closed jar and freezing.

Method

From Keim *et al.* 1988. Soybean Genetics Newsletter 15: 150-153 and Saghai-Maroof *et al.* 1984. PNAS 81:8014-8019.

1. Grind 1 gram dried (air-dried or freeze-dried) pea shoots to a fine powder in a mortar with pestle.

2. Mix the powder with 25 ml EB (extraction buffer) in a 50 ml, capped centrifuge tube.

3. Place the tube into a 65°C water bath and incubate it there for 1 hr. Several times during the hour, mix the tube's contents by inversion.

4. Remove the tube from the water bath and allow it to cool for several minutes on the bench. (Note: Do not skip this step. The chloroform added in the next step will boil out of the tube if added at 65°C.)

5. Add 20 ml chloroform to the tube, cap and mix by inversion until the contents are thoroughly mixed. When mixed, the extraction buffer and the chloroform will form a thick emulsion.

6. Centrifuge the tube at > 3500 x g for 10 minutes to break the emulsion and separate the tube contents into two phases.

7. Upon removal from the centrifuge, the contents of the tube form three distinct layers. At the bottom is a green layer of chloroform. Often, bits of plant material are pelleted under this layer. In the middle is an interphase consisting largely of denatured protein and bits of leaf tissue. This layer is whitish or yellowish in color. At the top of the tube is the straw-colored aqueous layer. This top layer contains the majority of the DNA in the preparation. Using a pipette, remove this top layer to a small flask or beaker. Avoid transferring any of the interphase material from the centrifuge tube.

Method, continued

8. What remains in the centrifuge tube, that is, interphase and organic (chloroform) layers are hazardous waste. Follow your instructor's direction for proper disposal.

9. To the aqueous phase in the flask or beaker, add 2/3 volume of isopropyl alcohol (for example, if you have transferred 24 ml to the flask, add 2/3 x 24 or 16 ml of isopropyl alcohol). Mix by swirling the contents. DNA will precipitate to from a cottony mass.

10. Using a glass rod or a pasteur pipet, transfer the DNA to a clean flask. Add 10 ml of 80 % ethanol, 15 mM ammonium acetate and swirl to wash.

11. After about 20 minutes, transfer the precipitated DNA to a microcentrifuge tube and centrifuge briefly to drive the DNA to the bottom of the tube. Using a pasteur or capillary pipet, remove the residual ethanol. Allow the DNA to dry in the uncapped tube for about 10 minutes on the bench top.

12. Add 0.75 ml of TE buffer to the DNA to dissolve the precipitate. (Note: Large quantities of DNA may require some time to dissolve completely. Leave the capped tube in the refrigerator until the next lab class.)

Day Two:

Cut 2 μl of the DNA with a restriction endonuclease as directed by your instructor.

On an agarose gel, compare total DNA (both uncut and cut) with DNA isolated from chloroplasts (both uncut and cut).

Results

The protocol is generally trouble-free. The only difficulties we have experienced with it come from: 1) using incompletely dried material, and 2) using heat-dried material. Rarely, DNA prepared by this method forms a flocculent rather than cottony precipitate. Such precipitates will not spool on a glass rod and must be collected by centrifugation. If desired, RNA can be removed from the DNA preparation by adding 20 micrograms of RNase A (heat-treated to destroy DNases) along with the TE buffer in Step 12 of the protocol.

Note on the use of chloroform in this exercise: If you want to avoid the use of chloroform in this exercise, make the following modifications to the lab protocol:

1. Grind tissue as before, mix with EB, and incubate in the water bath.

2. Centrifuge to pellet undigested plant tissue.

3. Transfer the supernatant to a fresh tube, flask or beaker and precipitate DNA with isopropanol as before. Note: The supernatant will be colored dark green in this preparation making the precipitation of DNA somewhat more difficult to see. If this is a problem, the supernatant can be diluted with additional EB or TE buffer before the precipitation step. DNA prepared in this way should not be expected to cut well with restriction endonucleases, but it may be satisfactory.

UNIT 3: CELL FRACTIONATION IN PLANTS

MODULE 9: ISOLATION OF CHLOROPLAST DNA

Mark Holland

* Introduction
* Safety Guidelines
* Experimental Outline
* Materials
* Pre-lab Preparation
* Method
* Results

Introduction

The DNA of plant cells is found in three distinct genomes. First, there is nuclear DNA, familiar as the DNA that makes up the chromosomes. But mitochondria and chloroplasts each have DNAs of their own. These genomes are closed circular DNA molecules encoding many of the enzymes necessary for the function of the organelles. Because of the importance of mitochondria and chloroplasts to the cell, their DNA is of interest to molecular biologists and biotechnologists. The chloroplast DNA of several species of plants have been cloned and sequenced in their entirety. In at least one organism (the green alga Chlorella), chloroplast as well as nuclear genomes have been genetically transformed. In this laboratory exercise, DNA will be isolated from chloroplasts and compared with total DNA. The objectives of this exercise are:

1. to isolate DNA from chloroplasts, and
2. to compare chloroplast DNA to genomic DNA.

Safety Guidelines

Some reagents used in this protocol are potential hazards. CTAB, cetyl-trimethylammonium bromide, is a strong detergent and can cause burns to the skin. Chloroform is toxic by inhalation or contact with skin. Follow instructor's direction for proper handling/ disposal of these materials.

Experimental Outline

Day 1 DNA isolation (2 hrs.)
- Add EB buffer to chloroplast prep
- Incubate 1 hr, 65°C
- Extract with chloroform
- Precipitate DNA
- Resuspend in Buffer

Day 2 Compare DNA from chloroplast to genomic DNA.
- Restriction Digest of DNA (optional)
- Pour gel, load cut and uncut DNA's

Materials

- EB (extraction buffer): 50 mM Tris pH 8.0, 1% CTAB, 50 mM EDTA, 1 mM 1,10-O-phenanthroline, 0.7 M NaCl, 0.1% beta-mercaptoethanol
- Chloroform
- Isopropyl alcohol
- TE buffer: 10 mM Tris pH 8.0, 1 mM EDTA
- Centrifuge Tubes
- Water Bath (65°C)
- Centrifuge

A kit, which contains all the critical reagents required for this experiment, is available from EDVOTEK (1-800-EDVOTEK).

Pre-lab Preparation

Chloroplasts are from Module 7 in this unit. Extraction Buffer and reagents are from Module 8 in this unit. No special preparation beyond the requirements for those exercises is required for this activity.

Method

1. Start with the frozen chloroplast preparation from Module 7. Typically, this sample will have a volume of several milliliters. For each milliliter of chloroplasts, add 4 ml EB. If necessary, transfer the mixture to a capped centrifuge tube of at least twice the volume of the chloroplasts and EB.

2. Incubate the mixture at 65°C for 1 hr.

3. Remove the tube from the water bath and allow to cool on the bench top for several minutes before proceeding.

4. Add an approximately equal volume of chloroform to the tube, recap and mix by inversion.

5. Centrifuge the tube at > 3500 x g for 10 minutes.

6. Upon its removal from the centrifuge, the tube contents will have separated into two distinct layers. Using a pipette, transfer the upper (aqueous) layer into a fresh centrifuge tube. (This tube should be of the same size as that used in the first step.) The lower (organic) layer is hazardous waste. Follow your instructor's directions for proper disposal.

7. Add 0.6 ml of isopropanol for each ml of DNA-containing extract in the centrifuge tube. Mix by inversion.

8. Centrifuge at > 10,000 x g for 20 minutes.

9. After centrifugation, decant the liquid in the tube away from the DNA-containing pellet. Stand the tube upside down on a paper towel or "Kimwipe" for several minutes to allow the liquid to drain. The tube's inside can be wiped carefully to remove liquid, but take care not to dislodge the DNA pellet.

Results

The yield of chloroplast DNA is expected to be low and will depend in some measure on the quality of the chloroplast preparation produced by the students in Module 7. Also note that chloroplast DNA may not cut well with restriction endonucleases. Still, it is likely that at least one lab group or individual will get results that can be shared with the rest of the class. Some of the students will accept this activity as a challenge. Since it is teamed with an examination of total DNA, none of the students will be completely without results. The exercise is worth a try if for no other reason than it will give the students some experience in working with very small quantities of DNA.

UNIT 3: CELL FRACTIONATION IN PLANTS

MODULE 10: ISOLATION OF MITOCHONDRIA AND ASSAY OF A MARKER ENZYME

Mark Holland

* Introduction
* Safety Guidelines
* Experimental Outline
* Materials
* Pre-lab Preparation
* Method
* Results

Introduction

In the following experiments, you will isolate mitochondria from the roots of pea seedlings by differential centrifugation and assay the activity of a mitochondrial enzyme to assess the success of the isolation protocol. Recall that the technique of differential centrifugation separates cell components based on differences in the rate at which they sediment in a centrifugal field. In the protocol used here, a first centrifugation step will remove whole cells, cell wall fragments, nuclei, starch, etc. Mitochondria, because of their small size will not be pelleted by this step, but will remain suspended in the supernatant. A subsequent spin at a higher speed will then be used to pellet the mitochondria to the bottom of the centrifuge tube.

To assess the efficiency of the isolation protocol, you will assay the activity of a “marker enzyme” in the mitochondrial pellet and in the supernatant from which the mitochondria are isolated. A marker enzyme is any enzyme whose activity is confined to the organelle being isolated. If the isolation protocol is 100% effective, all marker enzyme activity should appear in the mitochondrial fraction. The marker enzyme to be assayed in this experiment is cytochrome c oxidase.

Enzyme activity is measured by determining the rate of oxidation of cytochrome c. This can be followed, using a spectrophotometer, by an increase in absorbance at 550 nm by the reaction mixture during the initial, linear phase of the reaction.

Safety Guidelines

Follow standard laboratory safety procedures.

Experimental Outline

Harvest root tissue
Prepare extract
Isolate mitochondria by differential centrifugation
Assay mitochondrial activity

Materials

Method 1 & 2

Pea seedlings
Ice bucket
Homogenization buffer
- 70 mM sucrose
- 220 mM mannitol
- 0.5 g/l Bovine Serum Albumin
- 2.0 mM HEPES pH 7.4

Cheesecloth/Miracloth
Centrifuge tubes
Pasteur pipets
Small paint brush
Blender
Centrifuge (super speed)
Spectrophotometer

Method 3

Pea seedlings (grow seedlings as for regular protocol, but use whole seedlings for this experiment)
Homogenization buffer (same as in regular protocol)
Cheesecloth/Miracloth
Centrifuge tubes (4 per lab group)
Test tubes (3 per lab group)
Pasteur pipets
Dropper bottles of:
- Methylene blue
- Janus green
- Iodine (KI)
- Vegetable or mineral oil

Clinical Centrifuge
Blender
37°C Water Bath

A kit, which contains all the critical reagents required for this experiment, is available from EDVOTEK (1-800-EDVOTEK).

Pre-lab Preparation

Equipment Requirements

The protocol described here is written to be used with a refrigerated, super-speed centrifuge (e.g. Sorvall RC series) equipped with a fixed angle rotor (e.g. Sorvall SS34).

Other comparable centrifuges will perform equally well, but specified speeds and times of centrifugation might have to be changed slightly. The operations manuals for your particular centrifuge and rotor will explain these changes. If a super-speed centrifuge is not available, consider the alternative protocol for cell fractionation described below.

The assay of mitochondrial marker enzyme activity requires the use of a spectrophotometer with a band width of 5 nm or less. Before using the cytochrome c solution in the assay, check to see that it is completely reduced. To do this, add 1 or 2 crystals of sodium dithionate to 1 ml of cytochrome c stock solution. Transfer solution to the spectrophotometer cuvette and measure its absorbance at 550 nm and 565 nm. The A_{550}/A_{565} ratio should be 9-10.

A blender is required for tissue homogenization. A standard household blender will work well, but the blades should be as sharp as possible (some laboratories have been known to modify their blenders by replacing the standard blades with razor blades!).

Plant Material

The protocols in this unit all call for the use of pea seedlings as starting material. The seed is relatively inexpensive and easy to grow. For best results, use fresh seed. Seedlings should be about seven days old for the lab. To plant, soak seed overnight in a large container. Sow the seeds on a layer of about 1.5 inches of wet horticultural grade vermiculite in a standard nursery flat (21 x 10 x 2 inches). The seed can be sown thickly - nearly touching one another. (We usually use a 500 ml beaker-full of dry seed to plant one flat. At harvest, you can expect about 300 g of root material.) Cover the seed with 0.5-1 inch of vermiculite and water well. Cover the flat with plastic wrap to hold in moisture until the seedlings begin to emerge. Once the seedlings have emerged and the plastic has been removed, keep well-watered. The seedlings can be grown in the lab on a window sill, in a growth chamber or in the greenhouse. Shoot material, not used in this exercise should be saved, air dried or freeze dried, and stored for genomic DNA isolation (Module 3).

Pre-lab Preparation, continued

Solutions Required:

Homogenization Buffer
- 70 mM sucrose
- 220 mM mannitol
- 0.5 g/l Bovine Serum Albumin
- 2.0 mM HEPES pH 7.4

0.1M potassium phosphate buffer, pH 7.4

0.8M ascorbic acid

4% triton X-100

5 mg/ml cytochrome c

sodium dithionate crystals

Method

Method 1: Isolation of Mitochondria
(Adapted from a method of V. Peterson, University of Missouri)

All solutions should be kept ice-cold. Keep cell homogenate on ice while you are working.

1. Harvest 5 grams of 7-day-old pea roots, shake off vermiculite in which they are growing, and rinse in a beaker of distilled water.

2. Chop the roots into small pieces with a razor blade or scissors and put into a chilled blender with 20 ml ice-cold homogenization buffer. (Note: More than one 5 gram batch of roots can be homogenized at a time. If several lab groups are sharing the blender, they should use 20 ml of buffer for each 5 grams of roots homogenized, then divide the homogenate.

3. Homogenize the tissue with five 2-3 second bursts of the blender at high speed.

4. Filter the homogenate through four layers of cheesecloth plus one layer of Miracloth. It may be necessary to squeeze the filtrate through the cloth. Wear gloves.

5. Pour the filtrate into a centrifuge tube, balance against a tube of water or against a tube from another lab group, and centrifuge at 4°C , 700 x g, 10 minutes (2500 rpm in a Sorvall SS34 rotor).

Method, continued

6. Decant the supernatant into a clean centrifuge tube and centrifuge at 4°C, 10,000 x g, 10 minutes (9500 rpm in a Sorvall SS34 rotor).

7. Decant the supernatant from the tube into a beaker and save it on ice. You will use it for the next experiment. The pellet at the bottom of the centrifuge tube should contain isolated mitochondria. Wash them by gently resuspending them in 20 ml of fresh homogenization buffer. This is most easily done by pipetting 1-2 ml of the buffer into the tube and using a small paint brush to break up the pellet. Once the pellet is resuspended in this small volume, it can be diluted with the remaining 18-19 ml of buffer.

8. Recentrifuge the washed mitochondria as in step 6 above.

9. Discard the supernatant from this spin and resuspend the mitochondrial pellet in 5 ml of homogenization buffer.

Method 2: Assay of a Mitochondrial Marker Enzyme
(Adapted from a method of Doug Randall and Nancy David, University of Missouri)

1. From the ice bucket, remove 0.2 ml of the mitochondria preparation to each of two small test tubes and 0.2 ml of the 10,000 x g supernatant to one small test tube and allow them to come to room temperature on the lab bench. To one of the tubes containing mitochondria, add 0.8 ml of potassium phosphate buffer (a 1/5 dilution). To the other tube of mitochondria, add 1.8 ml buffer (a 1/10 dilution).

2. Assemble four reaction mixtures in four small test tubes:

Reaction Mixture (in each test tube)	
0.1 M potassium phosphate buffer, pH 7.4	50 µl
4% Triton X-100	25 µl
5 mg/ml cytochrome c	50 µl
Distilled water	675 µl
Sample to be assayed (mitochondria dilutions or supernatant)	200 µl

Method, continued

3. One of the four tubes above contains no sample to be assayed. Transfer the contents of this tube to a cuvette and use it as a blank in the spectrophotometer. If your spectrophotometer requires a reference cuvette, blank the cuvettes with water and use the contents of this tube as your reference.

4. To perform the assay, transfer a reaction mix containing sample to a cuvette and place into the spectrophotometer. Record the absorbance at 550 nm at 20 second intervals for one minute to determine the rate of the reaction in the absence of substrate. This value (the slope of this line) will be used in step 7 below as a correction factor in your calculation of the rate of the enzyme-catalyzed reaction. If your spectrophotometer uses a reference cuvette, this step is not necessary.

5. Start the reaction by removing the cuvette from the spectrophotometer and adding 10 µl of 0.8M ascorbic acid. Mix contents of the cuvette by inversion and quickly replace the cuvette in the spectrophotometer. Record the absorbance at 550 nm at 20 second intervals for 2 minutes. (Note: Check this time course.)

6. Repeat steps 4 and 5 for each sample to be assayed.

7. Calculate the rate of cytochrome c oxidation by each of the samples:

   $$\text{Rate} = \frac{\text{change in absorbance/minute (step 5)} - \text{change in absorbance/minute (step 4)}}{\text{molar absorptivity of cytochrome c} \times \text{path length of light through the cuvette}}$$

   Molar absorptivity of cytochrome c = $18.5 \times 10^6\ M^{-1}$
   Path length through cuvette is usually 1 cm.

The units attached to rate are moles of cytochrome c oxidized per minute. In general, this value would be reported as micromoles cytochrome c oxidized per minute, thus, calculation is simplified to:

$$\text{Rate} = \frac{\text{change in absorbance/minute (corrected as above)}}{18.5}$$

Note: Enzyme activity is usually reported in the literature as specific activity. This is a way of standardizing the reporting of enzyme activity since the exact molar concentration of enzyme in the preparation being assayed is not known. How is specific activity related to the rate of reaction just calculated? Specific activity = rate/ mg protein (total) in the assay mixture.

Method, continued

Method 3: Protocol Using Clinical Centrifuge Only
(Adapted from a method of A. Brown, Seattle Central Community College, Seattle, WA)

Objectives:

1. Separate a cell homogenate into several fractions by differential centrifugation.
2. Compare the composition of the fractions by microscopy.
3. Assay mitochondrial activity in the fractions by a qualitative assay.

All solutions should be ice-cold. Keep samples on ice at the lab bench if possible.

1. Harvest whole pea seedlings and wash in distilled water to remove planting material. Blot dry with paper towels.

2. Weigh out about 50 grams of seedlings, chop them into small pieces with a razor blade or scissors and transfer them to a chilled blender. Add 250 ml ice-cold homogenization buffer.

3. Homogenize tissue with five, 2-3 second burst of the blender at high speed.

4. Divide homogenate among lab groups. Each group should have 20-25 ml (enough for two, 10-15 ml centrifuge tubes).

5. Filter the homogenate through four layers of cheesecloth plus one layer of miracloth into a beaker on ice. Gently squeeze the cloth to remove most of the liquid. Wear gloves. Save the residue in the cheesecloth for examination later. Note: Homogenization of tissue and filtration step can be done prior to class by the instructor.

6. Divide filtrate in 2 centrifuge tubes. Tubes should be filled to same level to keep centrifuge balanced. Label tubes 1 and 2.

7. Centrifuge for three minutes at 200 x g . Start timing when the centrifuge rotor reaches top speed. Allow the rotor to coast to a stop when three minutes are up.

8. Remove the tubes from the rotor. Decant the supernatant from tube 2 into a clean tube (label the new tube 3). Save the pellet in tube 2 on ice. Add buffer to tube 3 to balance it against tube 1.

Method, continued

9. Return the tubes (1 and 3) to the centrifuge and spin at 700 x g for 10 minutes.

10. Remove the tubes from the rotor.

11. Decant the supernatant from tube 3 into a fresh tube (label 4). Save the pelleted material in tube 3.

12. Prepare wet-mount slides of the cell fractions (residue in cheesecloth and tubes 1-4) and examine them on the microscope. The residue in the cheesecloth consists largely of unbroken pieces of seedling tissue, cell debris, etc. Tube 2 contains material pelleted at low speed (200 x g) from the homogenate. Can you identify any of these components? Add a drop of iodine solution to the edge of the coverslip of your slide. Any starch grains present should turn blue-black in the presence of iodine. The pellet in tube 3 was separated from the homogenate at higher speed (700 x g) and for a longer time (10 minutes) than that in tube 2. How is this pellet different from the pellet in tube 2? Are any of the components present in these two pellets the same? The supernatant in tube 4 should be relatively clear. Only the smallest cell components remain in this solution. Prepare a wet-mount slide of this supernatant. Add a drop of Janus green stain. With the stain, at high magnification, mitochondria appear as tiny, dark specks. Examine tube 1. This tube shows a history of the entire experiment and illustrates the principle behind differential centrifugation. The sediments at the bottom of the tube are arranged in layers. The largest, most dense cell components are at the tube's bottom. Additional layers of sediment were added to the pellet by higher centrifugal forces applied for longer periods of time. Only the tiniest of particles remain in solution at the end of the experiment.

13. Assay the activity of mitochondria. To demonstrate that mitochondria remain in the supernatant of the high speed spin (tube 4) and that they have been separated efficiently from other cell components, perform the following experiment:

As you know, mitochondria are the subcellular sites of respiration. The activities of these organelles thus consume oxygen. Methylene blue dye is blue in the presence of oxygen, but is colorless when reduced (i.e. when oxygen is removed from it). In this experiment, the presence of mitochondria is detected by the disappearance of blue color from the reaction mixture. Prepare 3 test tubes according to the chart which follows:

Method, continued

Component	Reaction 1 (Control)	Reaction 2	Reaction 3
Buffer	6 ml	3 ml	3 ml
Pellet in tube 3 (resuspended in buffer)	—	3 ml	—
Supernatant from tube 4	—	—	3 ml
Methylene Blue*	2-3 drops	2-3 drops	2-3 drops

*the same amount should be added to each of the tubes.

Mix each of the tubes well, then add 1 ml of vegetable or mineral oil to the top of each. Incubate the tubes in a 37°C water bath several hours to overnight. Compare the tubes and record your observations.

Results

The experiment is generally very reliable. Use of pea roots rather than shoots avoids possible problems with chloroplast contamination.

Although different portions of pea seedlings are used in modules 6 and 8 of this unit, don't be tempted to save unused tissue from one experiment for the other experiment unless it can be used immediately. Fresh tissue gives the best results. Seedlings should be about seven days old (in any case, not older than twelve days).

Yield of mitochondria is difficult to predict and depends on a number of factors including the effectiveness of homogenization. For this reason, two dilutions of the mitochondrial fraction are assayed for marker enzyme activity.

UNIT 4

Animal Cell Culture

UNIT 4: ANIMAL CELL CULTURE

MODULE 11: ESTABLISHMENT OF A PRIMARY CELL LINE

Malethu T. Mathew

* Introduction
* Materials
* Safety Guidelines
* Experimental Outline
* Method
* Results

Introduction

The objective of this module is to establish primary cell cultures of chicken embryo fibroblasts. Primary cell cultures are those that are composed of cells taken directly from a living animal. Primary cells normally contain a diploid set of chromosomes, have limited life-span and undergo aging. In order to prepare a primary culture, an organ or tissue is removed from a freshly sacrificed animal and aseptically cut into small pieces and treated with enzymes such as trypsin, collagenase, pronase or combination thereof. After the cells have been separated from the tissue, they are inoculated into appropriate cell culture medium in tissue culture flasks. The cells will adhere to the surface of the flask and replicate until they come in contact with each other. Thus they attach and grow as a uniform layer of cells, or a monolayer, which is always one cell thick. These cells will stop growing once the surface is filled up and contact follows. This is known as contact inhibition. Cells can be cultured as stationary monolayers usually inoculated with 2 x 10^5 cells in 5 ml of cell culture medium in a petri dish designed for tissue culture work.

Some cell types can be grown in suspended state in culture medium. Suspended cells are generally derived from blood cells. B lymphocytes can be established in culture directly from leukemic cells or by infecting normal lymphocytes with Epstein-Barr virus. Suspension cell cultures will yield 5-10 times more cells per ml of medium than monolayer cultures.

It is fairly easy to derive a primary culture of fast growing cells like fibroblasts, and as a result most primary cultures consist of fibroblasts. Fibroblasts are connective tissue cells which secrete the extracellular matrix of connective tissue. In this exercise, primary fibroblast cultures will be derived from chicken embryo. They will be made by dissociating the entire embryo with the proteolytic enzyme trypsin. Embryo cells are somewhat easier to adapt to cell culture than adult cells, since embryo cells are less differentiated and more likely to be rapidly dividing. The culture that is first produced will contain a variety of cell types. It is possible, however, to eventually establish a homogeneous culture of fibroblasts, because the other cell types do not grow as fast and they will be diluted out in subsequent passages subcultures.

Safety Guidelines

Wear gloves, lab coat, and safety glasses. Standard laboratory safety procedures should be followed.

Experimental Outline

- Establish primary cell culture from chicken embryo
- Culture cells as stationary monolayer in tissue culture petri dishes
- Establish a homogeneous culture of fibroblasts
- Grow cells in tissue culture medium under sterile conditions
- Stain cells with trypan blue and count the number of unstained, live cells under a microscope using a hemocytometer

Materials

fertile chicken eggs, incubated 7-10 days
a bottle of 0.25% trypsin solution
a bottle of PBS (phosphate buffered saline)
tissue culture flask 25 cm^2
a bottle of HMEM (Hank's minimum essential medium) + supplemental 10% Fetal Bovine Serum (heat inactivated)
three 100 mm plastic petri dishes
one 50 ml plastic beaker
150 ml squeeze bottle with 70% alcohol
a sterile dissection kit containing sharp scissors, sharp scalpel and two pointed forceps
95% ethanol
5 ml plastic syringe
sterile gauze
two 125 ml flasks
trypsinizing flask
15 ml centrifuge tubes (2)
trypan blue (0.4%)
0.15 ml plastic microcentrifuge tube with an attached cap
penicillin-streptomycin 100x stock solution
1 ml and 5 ml pipets, sterile
pipet pumps
CO_2 incubator

Method

1. Keep 0.25% trypsin, HMEM and PBS in 37°C water bath.

2. Aseptically place 15 ml of PBS into each sterile petri dish.

4. Place the egg in a 50 ml beaker with its blunt end up, and disinfect the entire egg shell with 70% ethanol.

5. With sharp sterile forceps puncture the top of the shell and remove the shell all the way up to the base of the air cell.

6. Locate the position of the embryo and carefully tear the membrane away from the embryo with sterile forceps.

7. Insert a sterile pair of forceps into egg and grab embryo firmly. Take it out of egg and place in the first dish of PBS.

8. Cut off the head and feet of embryo with a sterile scissors. This procedure is done for 10 day old embryos but not for younger embryo.

9. With a flamed sterile forceps, transfer the remaining parts of the embryo to a petri dish containing PBS and rinse thoroughly. Repeat the rinse until no color is found in the rinse solution in the dish.

10. Take a sterile 5 ml syringe and a sterile capped test tube. Remove the plunger from the syringe and rest it on sterile kimwipes. Use a sterile forceps to transfer the remaining parts of the embryo from the petri dish into the syringe. Remove the cap from the tip of the syringe without touching the tip with your fingers. Uncap the sterile test tube and place the cap on sterile kimwipe and keep the tube in a test tube rack. Position the tip of the syringe over the test tube. After inserting the plunger back in the syringe, with a steady and even pressure, push the plunger all the way down so that the embryo is forced through the syringe into the test tube. The embryonic tissues are delicate and are homogenized to very fine pieces.

11. Transfer the tissue into a trypsinizing flask and add 10 ml of 0.25% prewarmed trypsin solution.

13. Place the trypsinizing flask on the shaker in the 37°C water bath and allow the suspension to swirl on the shaker at low speed so that the tissues of the embryo just barely graze the cutting edges of the trypsinizing flask. It will be necessary to continue this action for 15-60 minutes depending on the disaggregation process, and you will have to monitor this process closely. The flask is removed once the trypsin solution becomes cloudy with individual cells. Excessive trypsinization will reduce the cell viability significantly.

Method, continued

14. Allow fragments to settle; the supernatant is a cloudy suspension of cells. Pour this cell suspension through the side spout into a sterile 125 ml flask containing 10 ml of prewarmed HMEM with 10% Fetal Bovine Serum. The serum will inhibit further action of trypsin and prevent further damage to cells.

15. Obtain another sterile 125 ml flask and place two or thee layers of sterile gauze over the mouth of the flask and filter the cell suspension through this gauze to remove large cell clumps and other debris.

16. Take two clean sterile centrifuge tubes. Aseptically transfer the cell suspension into these tubes so that both tubes contain the same quantity. Pellet the cells by centrifugation in a table top centrifuge for 5 minutes at 1000 rpm. Discard the supernatant and resuspend the cell pellets in 5 ml of HMEM containing 10% Fetal Bovine Serum. Combine the cell suspension in one tube and cap it.

17. Remove the cap of the tube containing the cell suspension and by using a sterile 10 ml pipette, aspirate the suspension by drawing it up into the pipette and then forcing back into the tube. Do this about six times. It is important to keep the cell suspension uniform and the clumps broken up as much as possible. Determine the viability and cell density.

18. With a sterile 1 ml pipette, aseptically transfer 0.1 ml of cell suspension into 1.5 ml microcentrifuge tube.

19. Pipette out 0.9 ml of 0.4% (w/v) trypan blue solution into the above microcentrifuge tube (wear gloves when working with trypan blue). Cap the tube tightly and vortex it for 15 seconds. Immediately with a Pasteur pipette introduce a drop of it at the center of one edge of the cover glass covering one of the counting chambers of the hemocytometer. The fluid will be drawn under the cover glass by capillary action. Do not over fill the chamber.

20. Place the counting chamber on a regular microscope and focus with low power objective on the ruled area of the chamber. The hemocytometer is divided into 9 large blocks (1 mm^2), and each of these large blocks is divided into 16 smaller squares. Count the number of unstained cells (trypan blue penetrates damaged cells, and thus only stains those cells that are dead) in the central large block and four large corner blocks. If there are too many cells to count, make the necessary dilution.

Method, continued

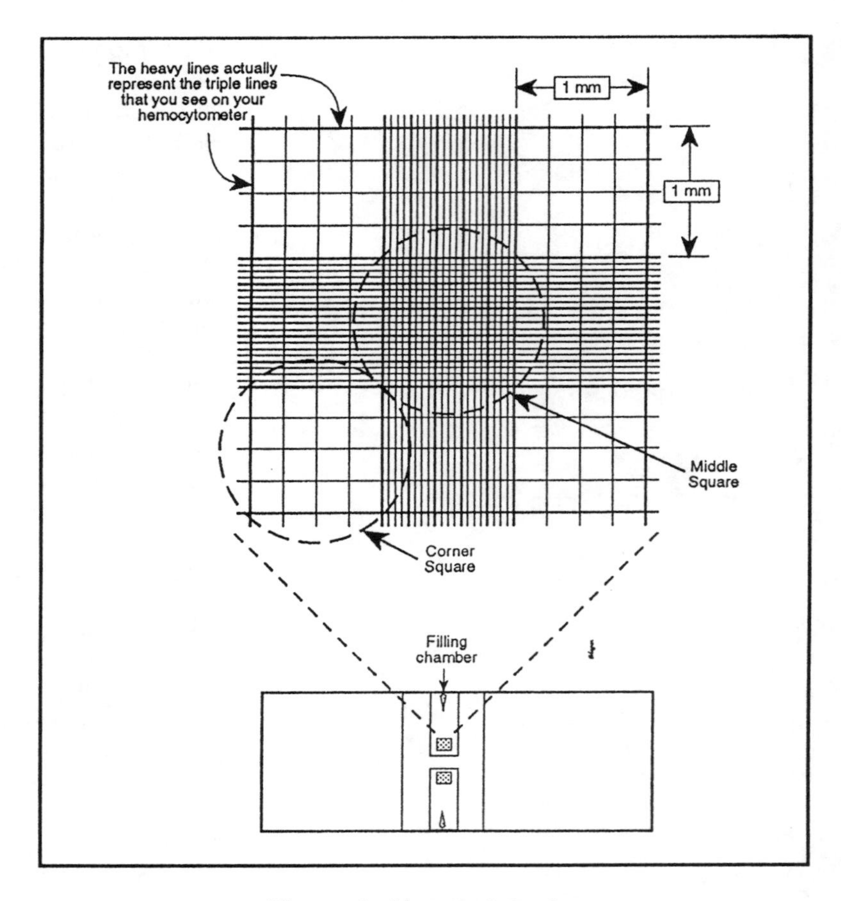

Figure 1. Hemocytometer

21. Divide total number of cells by 5 to obtain average number of cells per large block. (Note: if this number is smaller than 10 and larger than 100, the count will not be accurate.) Multiply by 10^4 and the dilution factor to obtain number of cells per ml in the suspension. An example is shown below:

Total cells counted in give large blocks	=	250
Average number of cells/block	=	50
Number of cells per ml	=	50×10^4

For the correct count, multiply the above number by 10 (dilution factor). The correct number of cells would be:

50×10^5

5×10^6

22. Adjust the cell density to 5×10^5 cells/ml by proper dilution. Aseptically transfer 5 ml of the diluted cell suspension into each of two 25 cm^2 tissue culture flasks.

Method, continued

23. Label each flask with your group name, date and place in incubator at 37°C Humidity can be provided by placing a pan of water in the incubator.

24. After 12-24 hours observe the cultures, pour off the old medium and replace it with 5 ml of fresh HMEM supplemented with 10% Fetal Bovine Serum. Change the medium every 2-3 days until the cells have become confluent. Notice how culture changes over time into homogenous line of fibroblasts.

Results

Count the number of cells and record this number.

UNIT 4: ANIMAL CELL CULTURE

MODULE 12: EXAMINATION AND MAINTENANCE OF A CELL LINE

Malethu T. Mathew

* Introduction
* Safety Guidelines
* Experimental Outline
* Materials
* Method
* Results

Introduction

The objective of this module is to culture and maintain established cell lines. A primary culture is split to produce new cultures and is subsequently known as a cell line. The two types of cell lines are (a) finite cell line and (b) continuous or established cell lines. A finite cell line is capable of a limited number of cell generations in vitro and after which cells die. A continuous cell line has the capacity for an infinite number of population doubling and thus in a sense immortal. Some continuous cell lines are also malignant, meaning they will grow as a tumor and invade other tissues if injected into an animal.

Earle is credited with establishing the first immortal cell lines in 1940. Most of these were established by treating rodent primary culture with chemical agents. One of these lines called L-cells was established in 1943 by treating primary mouse fibroblasts with the cancer causing agent methylcholanthrene. George and Margaret Gey derived cells from the cervical carcinoma of a black woman named Henrietta Lacks. This cell line which is called "HeLa", is the most common source of human cells used in research today. Today, many cell lines are available which can be obtained from American Type Culture Collection (ATCC), local medical colleges or any research institutions.

Established cell lines can be divided into two groups: those that can grow attached to a solid surface (monolayer) and those that grow in suspension. The mouse L-cells is a good example of the surface dependent cell line. When these cells are seeded in a culture vessel, they attach and grow to form a uniform layer of cells or a monolayer, which is always one cell thick. These cells will stop growing once they start filling up the surface and touching each other due to contact inhibition. In order to subculture a monolayer, it has to be dissociated to individual cells with trypsin, after which it can be put in a larger vessel with more space and fresh medium. Suspension cells are grown in liquid culture similar to bacterial culture.

Safety Guidelines

Wear gloves, lab coat, and safety glasses. Standard laboratory safety procedures should be followed.

Experimental Outline

- Maintain an established cell line in tissue culture
- Introduce Trypsin-EDTA into the culture flask, aseptically to help maintain conditions for a monolayer culture
- Observe individual cells using an inverted microscope
- Maintain cells, replace medium when a drop in the pH is observed based on change on phenol red indicator - changing from red to yellow

Materials

chick embryo fibroblasts (from Module 11) or mouse L-cells

a bottle of PBS

a bottle of 1x Trypsin-EDTA solution (concentrated)

trypan blue solution - 0.5% in PBS

a bottle of complete medium (HMEM), supplemented at 10% v/v with Fetal Bovine Serum and 1% v/v penicillin-streptomycin 100x stock

waste container

10 ml pipets, sterile and pipet pump

CO_2 incubator

Method

1. Keep PBS, Trypsin-EDTA and complete medium in a 37°C water bath.

2. Observe the condition of the monolayer by holding the bottle up to the light. In a good culture the medium will appear clear and the cells will look smooth and confluent with no clumps.

3, Observe the individual cells by using an inverted microscope at low power. Only those cells which undergo mitosis would be round and floating.

4. Make sure that the monolayer is confluent and if so, it is ready to be transferred.

5. Uncap culture flask and decant medium to a waste container.

6. Aseptically withdraw 10 ml of sterile PBS, and transfer it into the culture flask and recap the flask. Rock the flask gently from side to side for 10-15 seconds to rinse the cells. This step helps to remove the serum, which would inhibit the enzymatic activity of trypsin. Decant the solution into a waste container.

7. Add 0.2 to 0.5 ml of Trypsin-EDTA into the culture flask, swirl the trypsin around and incubate at 37°C for 3 minutes. Too much trypsin can damage cells. Cells are attached to cells and substrate by surface glycoproteins, serum proteins and chemical groups that coat the tissue culture-treated plastic surfaces of the flask. The enzyme trypsin catalyzes the hydrolysis of bonds involved in cell to cell and cell to substrate contact.

8. Once the cells begin to dislodge, tap the flask against the palm of your hand lightly. Hold the flask up to the light to check that no patches of monolayer remain. Observe the cells under an inverted microscope to confirm their presence.

9. Aseptically add 5 ml of complete medium. Proteins in the serum will stop the action of trypsin.

10. Aspirate the cell suspension by drawing it up into the pipette and expelling it back into the flask 5-6 times with the tip of the pipette pressed against the bottom corner of the flask. This will break up the cell clumps. Clumped cells will affect the accuracy of the cell count.

11. Count the cells using a hemacytometer. (Follow the procedure as described in Module 1.)

Method, continued

12. Expand the culture by taking 1 x 10^5/ml for chicken fibroblasts or 1 x 10^4/ml for L-cells. If you are using 25 cm^2 flasks, transfer about 1 ml of cell suspension to each flask and add 9 ml of medium to the flask.

13. Incubate the culture at 37°C for 12-24 hours. After this, decant the old medium and add 10 ml of fresh medium and incubate. It is necessary to transfer (split) these cultures once they become confluent, which should occur in 3-4 days. Observe them under inverted microscope every 2-3 days. When the phenol red indicator in the medium changes from red to orange-yellow or yellow which indicates a drop in pH, decant and discard the medium and replace it with fresh medium.

Results

Results can vary from group to group. Students can quantify viable cells when utilizing the inverted microscope.

UNIT 4: ANIMAL CELL CULTURE

MODULE 13: CELL POPULATION ASSAY

Malethu T. Mathew

* Introduction
* Safety Guidelines
* Materials
* Method
* Results

Introduction

The objective of this module is to prepare a growth curve for a given cell culture. An important growth parameter used for cultured cells is the population assay. It is a good index from which to establish the best culture conditions for a cell line. To determine a growth curve, cells are plated at low densities, and are counted at fixed time intervals, without feeding or diluting the culture. The number of cells in the culture is plotted against the elapsed time, and this data generates a curve which can be divided into four stages.

The first stage of a growth curve is called the lag phase. In this phase of growth, the cells grow slowly or not at all. At this stage cells adapt to new conditions such as type of medium, serum concentration, type of culture vessel and initial cell density. These factors determine the length of the lag phase.

The next stage of a growth is called log phase. Here the cells will grow exponentially and will divide as fast as possible. In this stage, the medium is rich in nutrients, and there is still sufficient space in the culture vessel available for the cells to grow without inhibiting or competing with each other. It is during the log phase that the doubling time of a culture should be determined. Doubling time is the time it takes for a culture in log phase to increase by a factor of two.

The next stage in growth curve is called the plateau phase. Here the cells begin to exhaust the available nutrients and growing space, and their rate of division slows down. At this stage, the maximum number of cells which can be grown per unit volume of medium can be determined. This saturation density represents the density at which the cells can no longer grow exponentially and, it too, will vary depending on the conditions used.

The last stage of the curve is called the death phase. With the depletion of nutrients and accumulation of waste products, the cells will begin to die.

Safety Guidelines

Wear gloves, lab coat, and safety glasses. Standard laboratory safety procedures should be followed.

Materials

a bottle of complete Fischer's medium (Fischer's medium supplemented with 10% v/v with horse serum, plus 1 mM sodium pyruvate)

culture of mouse L-5178Y cells or any line that grows in suspension (note that other cell lines may require different media)

concentrated trypan blue solution - 0.4%

CO_2 incubator or a tank of 5% CO_2

Method

1. Obtain a culture of L-5178Y cells growing in 25 cm^2 flask in Fisher's medium.

2. Tighten the cap of the flask and shake it well from side to side.

3. Using a 10 ml pipette, aspirate cell suspension up into and out of the pipette 10 times to break up the clumps of cells.

4. Transfer 0.1 ml of the cell suspension to a 1.5 ml microcentrifuge tube with attached cap. Close the culture flask with its cap.

5. Add 0.1 ml trypan blue solution and count the cells using a hemocytometer Trypan blue is used to check viability.

6. Record the cell count in the original cell suspension. Make proper dilution so that it now contains 2×10^4 cells/ml in about 20 ml of the medium. This is the culture that will be used for generating growth curve.

   Using the formula $V_1C_1 = V_2C_2$ calculate what volume of the original cell suspension (V_1) when diluted to a final volume of 20 ml (V_2) yields a final density of 2×10^4 cells/ml (C_2). If the original density (C_1) of your cell suspension is 1×10^6 cells/ml, you will set up the formula as follows:

$$V_1 (1 \times 10^6) = (20 \text{ ml}) (2 \times 10^4)$$

$$V_1 = \frac{(20 \text{ ml} \times 2 \times 10^4) \text{ cells}}{1 \times 10^6 \text{ cells}}$$

$$= \frac{40 \times 10^4}{1 \times 10^6} = \frac{4 \times 10^5}{1 \times 10^6} = \frac{4}{10} = 0.4 \text{ ml}$$

7. Transfer this volume into a graduated sterile cylinder and bring the volume to 20 ml using Fisher's medium. Transfer this to 25 cm^2 culture flask and cap it. This culture will be used for generating a growth curve. Count this culture at least twice a day.

8. Place the flasks upright in the CO_2 incubator at 37°C Make sure that the cap is loosened before returning the cells to the CO_2 incubator. Alternatively, gas the culture with 5% CO_2 and keep the cap tightened. Remember that the concentration of cells at 0 time is 2×10^4 cells/ml.

Method, continued

9. At least two measurements should be taken on the first day of culture. For example, at about three and nine hours after time zero are ideal.

10. Note the time of day when each set of daily measurements are taken and calculate the age of the culture, in hours from time zero, for each set of density determinations. Keep a written record of all cell counts. After each set of determinations, return the flasks to the CO_2 incubator, with caps loosened or to the regular incubator, with caps tightened following a 10-15 second bubbling of 5% CO_2.

11. Continue this procedure until the medium has turned yellow and the density of viable cells has declined considerably.

Results

Once the data has been collected, construct a graph using semilog paper by placing the time in hours on the x-axis and the cell count on the y-axis.

UNIT 5

Antibody Specificity and Diversity

UNIT 5: ANTIBODY SPECIFICITY AND DIVERSITY

MODULE 14: GEL DIFFUSION/OUCHTERLONY

Karen K. Klyczek

* Introduction
* Safety Guidelines
* Experimental Outline
* Materials
* Pre-lab Preparation
* Method
* Results

Introduction

The specific combination of an antibody (Ab) with an antigen (Ag) is the fundamental reaction of immunology. An antibody is a large protein molecule which contains two or more identical sites for combining with a specific antigen. A single antibody molecule can bind simultaneously to more than one molecule of antigen. Most antigens are also macromolecules such as proteins, and each antigen usually contains multiple binding sites and can be bound simultaneously by more than one antibody. Under appropriate conditions of antibody and antigen concentration, antibodies and antigens can form large complexes, or lattices, which are insoluble and precipitate from solution (Fig. 1). The ability to form precipitates makes it possible to perform qualitative and quantitative assays using antigen-antibody systems. Such precipitation assays are very sensitive and relatively quick and easy to perform.

Precipitation assays can be performed either in solution or in a gel such as agarose. When a protein is placed in a well which has been cut in an agarose gel, the protein will diffuse away from the well, through the gel, in all directions, to create a gradient of the molecule where the concentration decreases with increasing distance from the well. This occurs because proteins can move freely in the gel, which microscopically is more like a matrix with many holes than like a solid mass, and the kinetic energy of the molecules will cause them to move randomly. If antibodies and antigens are placed in wells in different areas of a gel, they will move toward each other, as well as all other directions, and will form opaque bands of precipitate (precipitin lines) where diffusion fronts meet (Fig. 2a). Maximum precipitation will occur in the area of gel where concentrations of antibody and antigen molecules are at an optimal ratio (equivalence zone; Fig. 1b). Closer to the antibody well, antibody concentration will be too high for precipitate formation (antibody excess zone; Fig. 1a). Closer to the antigen well, the antigen concentration will be too high (antigen excess zone; Fig. 1c).

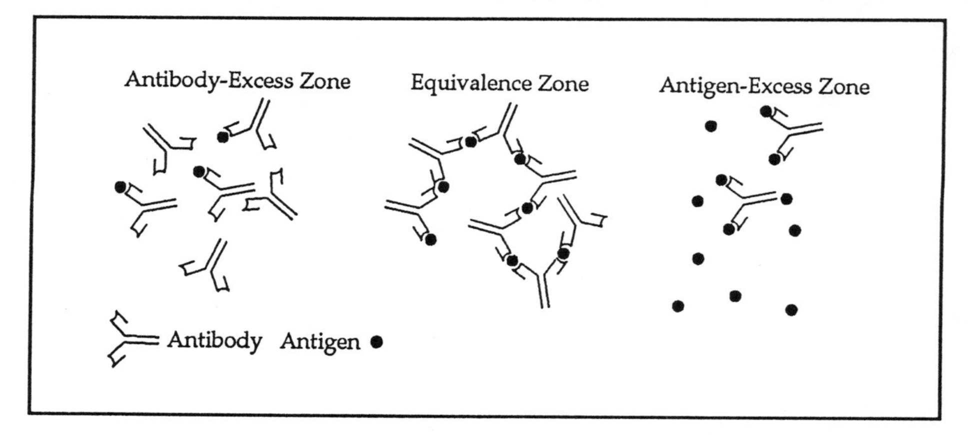

Figure 1. Importance of relative antibody and antigen concentration in formation of a precipitate.

Introduction, continued

A major advantage of this type of gel diffusion assay is that more than one antigen-antibody system in a mixture can be detected. It is important to remember that when antibodies are produced by injecting an antigen into an animal, many different antibody molecules reacting with different parts of the antigen (e.g. different proteins in a mixture, or antigenically different parts of a single protein) will be found in the serum of the animal. It is this serum, referred to as antiserum, which is usually used as a source of antibodies in experiments such as these. A single antigen will combine with the antibody it induces (its homologous antibody) to form a single precipitin line in a gel diffusion assay. When two antigens are present, each behaves independently of the other. The positions of the precipitin lines will depend primarily on the sizes of the antigens and their rates of diffusion in the gel. Thus, the number of precipitin lines indicates the number of antigen-antibody systems present. (Fig. 2b). This is actually the minimum number of antigen-antibody systems, since different precipitin lines often develop near each other and appear as a single line.

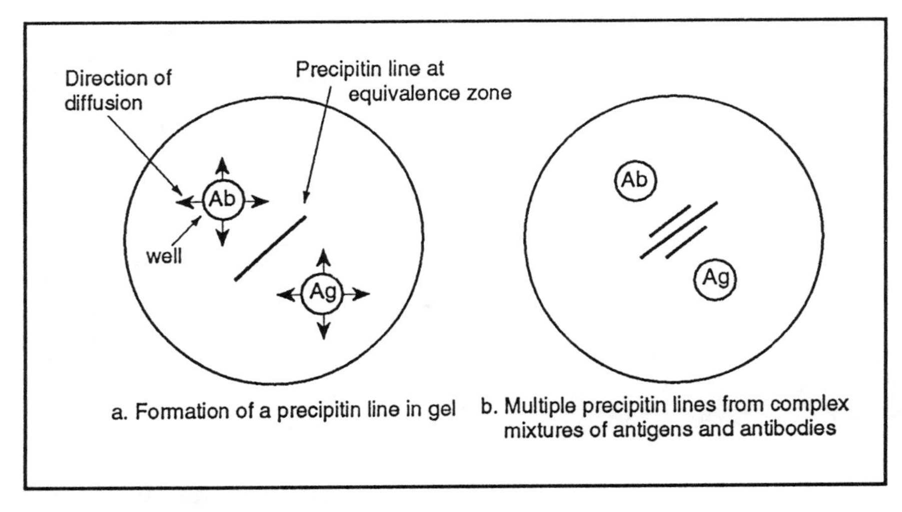

Figure 2. Diffusion of antigen and antibody and the formation of precipitin lines

Double diffusion, referring to the process described above in which both antigen and antibody molecules diffuse from a well, is a simple but very useful procedure invented by the Swedish scientist, Ouchterlony. It is used to compare antigens for the number of identical or cross-reacting determinants. If a solution of antigen is placed into two adjacent wells, and the antibody is placed equidistant from the two antigen wells, the two precipitin lines that form will join at their closest ends and fuse. This is known as a reaction of identity (Fig. 3a). When unrelated antigens are placed in adjacent wells and the center well is filled with antibody reacting with both

Introduction, continued

antigens, the precipitin lines will form independently of each other and will cross. This is known as a reaction of non-identity (Fig. 3b). If one of the antigens was used to produce the antibody sample (i.e. it is the homologous antigen), and the second antigen shares with the homologous antigen some, but not all, of the specificities recognized by the antibody (i.e. it is a cross-reacting antigen), the two precipitin lines will fuse. However, an additional spur will form which projects toward the cross-reacting antigen (Fig. 3c). The spur represents the reaction between the homologous antigen and those antibody molecules which do not bind to the cross-reacting antigen.

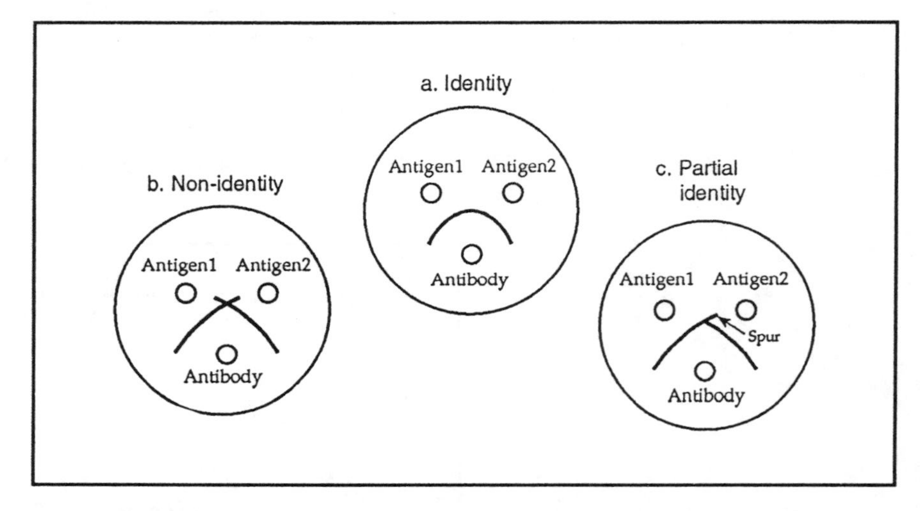

Figure 3. Relationships between antigens which can be demonstrated by the double diffusion procedure.

In this exercise, you will carry out a double diffusion assay using antibodies produced in goats against rabbit serum proteins. The proteins in the rabbit serum are recognized as foreign by the immune system of the goat, and antibodies are produced against the individual serum proteins. The relationships between different antigens, including the whole serum used to generate the antibodies and individual serum proteins such as albumin and immunoglobulin, will be explored. Goats and rabbits, as well as sheep, are commonly used in immunology research as sources of serum and antiserum, since they are relatively easy to maintain and yield larger volumes of serum than mice or rats, and they generally mount good antibody responses when injected with a variety of antigens.

It is important not to confuse the terms serum and antiserum as used in this experiment. Serum refers to blood with the blood cells and clotting proteins removed. Serum is useful as an antigen

Introduction, continued

sample because it contains many proteins which are antigenic when injected into a different species of animal. It is also relatively easy to isolate serum from an animal. Antiserum is serum from an animal which has been injected with an antigen (e.g. serum from an animal of a different species, hence the potential confusion), and therefore has produced antibodies against the antigen. These antibodies are found in the animal's serum, which is now called antiserum if it is used as a source of antibodies. Similarly, the terms immunoglobulin and antibody must be distinguished. Antibody molecules are immunoglobulin proteins. However, in this exercise, immunoglobulin means antibody molecules present in the serum of an animal not injected with a specific antigen. Immunoglobulin in the rabbit serum is one of the protein antigens present in the serum, and the goat injected with the rabbit serum will produce antibodies against the rabbit immunoglobulin.

Note that the Ouchterlony assay is a qualitative test, in contrast to radial immunodiffusion (RID; Module 19), which is quantitative. In an RID assay, a constant amount of antibody is mixed into the gel, and antigen is placed in a well cut into this gel. Diffusion is one-dimensional, since only the antigen diffuses. The position of the precipitin line which forms is used to determine the concentration of antigen in the test solution, with the aid of a standard curve.

Safety Guidelines

1. Use caution when melting agarose, particularly in a microwave. Melted gel may boil over top of flask. Protect fingers when handling the hot flask, and swirl hot flask very gently at first.

2. Wear gloves, lab coat, and safety glasses when working with Coomassie blue and destain. The destain solution contains methanol, which is toxic, and acetic acid, which is corrosive. In addition, the stain will stain skin and clothing if spilled.

3. Use caution when cleaning glass slides, since edges and corners may be sharp.

4. No food or drink should be used in the laboratory.

5. No other special safety or disposal measures are required, other than standard lab safety procedures.

Experimental Outline

Day 1: Set up experiment

1. Coat slides with melted agarose and allow to solidify. Then cut sample wells in agarose.

2. Place antigen and antibody samples in appropriate wells, and place slides in incubation chamber to allow precipitin lines to form.

Day 2: Analyze results

1. Observe and record appearance of precipitin lines

2. Interpret and compare with predicted results.

Optional Staining and Destaining:

1. Soak slides in Tris buffer.
2. Soak slides in staining solution.
3. Remove stain and soak slides in several changes of destain.

Materials

1% agarose, dissolved in buffered saline solution, pH 7.4, and kept in 60°C water bath
4 microscope slides
95% ethanol
5 ml pipet
pipet pump
well cutter
template for cutting wells (see step 5)
5 transfer pipets with bulb
practice loading solution (optional)
tube A, containing antibodies produced against whole rabbit serum (Antibody, or Ab)
tube B, containing whole rabbit serum (Antigen B, or AgB)
tube C, containing rabbit serum albumin (Antigen C, or AgC)
tube D, containing rabbit serum immunoglobulin (Antigen D, or AgD)

For Optional Staining and Destaining:

0.1 M Tris buffer, pH 7.5
50:40:10 water:methanol:acetic acid
0.1% Coomassie blue in 50:40:10 water:methanol:acetic acid

NOTE: A kit may be purchased from EDVOTEK, Inc., "Antigen-Antibody Interaction: The Ouchterlony Procedure" (EDVO-Kit #270), which provides all of the materials marked with an asterisk, and is sufficient for 10 groups.

*Antiserum produced in goats against rabbit serum, approximately 2 ml
*Rabbit serum, approximately 2 ml
*Rabbit albumin, 0.01 mg/ml, approximately 2 ml
*Rabbit immunoglobulin (IgG), 0.01 mg/ml, approximately 2 ml
*Phosphate-buffered saline (200 ml), either purchased or made (recipe below)
*Agarose, 1.8 g
250 ml flask
*Microscope slides (40)
*Transfer pipets (50), e.g. plastic pipets with self-contained bulb**
*Well cutter (10), such as cut pieces of drinking straws, or pasteur pipets with bottoms cut off where pipet begins to enlarge and fire-polished
Plastic or glass pan, with sufficient room for all experimental slides
Paper towels, to generate a 1/2 inch stack in pans
Plastic wrap or lid for pans
*5 ml pipets (10)
Ethanol in squirt bottles, for washing glass slides
Distilled water

Materials, continued

Pipet pumps or bulbs for 5 ml pipets (10)
Marking pen or grease pencil
Toothpicks or spatulas for removing agarose plugs from wells (10)
*Practice loading solution: any colored solution such as food coloring (optional)
Tris, 0.01 M, pH 7 (optional)
Staining solution: Coomassie blue, 0.1% in water:methanol:acetic acid 50:40:10 (optional)
Destain: water:methanol:acetic acid 50:40:10 (optional)

**A good alternative is microcapillary pipets, with 5 microliter graduations, with a plunger for drawing up and dispensing liquid. These are very easy to manipulate. Most will dispense a maximum of 25 microliters, which still should be sufficient to yield visible precipitin lines.

Pre-lab Preparation

Timetable of Events

Preparation of agarose-coated slides and loading of wells, up to the point of incubation, can be completed in one two-hour lab period. Observation and analysis of results can then be scheduled for the following lab period (one week later, if the slides are stored in the refrigerator) or during part of a class period in the next day or two. Staining of slides, if desired, is most easily done by the instructor once the precipitin lines have formed. The staining procedure takes 1-2 days and therefore would delay observation by the students. Alternatively, students could carry out the staining on their own, if they are willing to come in outside of class time.

Pre-lab Preparation

1% agarose:	20 min. (35 min. if making phosphate buffered saline)
Incubation chamber:	5 min.
Staining solutions (optional):	20 min.

Lab procedure

Preparation of slides:	25 min. (10 of this for solidification)
Cutting of wells:	15 min.
Loading samples:	15 min.
Incubation of experimental slides:	overnight at room temp, up to one week refrigerated

Pre-lab Preparation, continued

Dialysis, staining, destaining: (optional)	1-2 days total time (only brief, occasional time commitments)
Observation/recording of results:	15 min.

NOTE: The slides may be prepared several weeks in advance, carefully wrapped in plastic and stored in the refrigerator.

Preparation of 1% agarose

1. If you are making phosphate-buffered saline, dissolve the following in 150 ml distilled water:

   2.19 g NaCl
   1.28 g KH_2PO_4
   2.63 g $Na_2HPO_4 \cdot 7H_2O$

   Adjust pH to 7.2, add distilled water to bring final volume to 250 ml

2. 1% Agarose solution:

   a. Add 180 ml phosphate-buffered saline to 250 ml flask. Add 1.8 g agarose. Swirl to disperse large clumps.

   b. Heat to dissolve agarose using a hot plate or Bunsen burner to bring mixture to boiling with occasional careful swirling. Periodically remove flask from heat and check to see if the small, clear particles of agarose have dissolved. Continue heating until all particles have dissolved and the final solution is clear.

      A microwave oven can also be used by heating the flask on high for 2 minutes, or until liquid begins to boil. Swirl the flask, microwave for an additional 1-2 minutes, taking care not to let the agarose boil over. Check to see that all agarose particles have dissolved.

   c. Cool the agarose solution to 60°C with swirling to promote even dissipation of heat. Partially immersing the flask in water and swirling will greatly increase the rate of cooling.

   d. If desired, divide the melted agarose into smaller flasks to distribute to groups.

   e. Keep melted agarose in 60°C water bath to prevent solidification. If the agarose does solidify, remelt it in a hot water bath or in a microwave oven.

Pre-lab Preparation, continued

3. If desired, divide the antiserum and antigen samples into labeled tubes. Label tubes according to list of materials.
4. Prepare incubation chamber by lining the bottom of a plastic or glass dish with 1/2 inch of paper towels. Soak the paper towels with distilled water. There should not be any layer of liquid above the paper towels. All liquid should be absorbed into the paper toweling. Cover the chamber with plastic wrap or a lid, if the dish has one.

5. If staining of slides is desired, Tris solution for dialyzing unprecipitated proteins, stain, and destain will be needed. To make Tris solution (0.01 M, 1 liter), dissolve 1.21 g Tris base in about 700 ml distilled water. Adjust pH to 7.0 and add distilled water to bring the volume to 1000 ml.

   For stain and destain, make 1 liter destain solution by mixing 500 ml distilled water and 400 ml ethanol, then add 100 ml glacial acetic acid slowly while gently stirring. For stain, remove 200 ml of this solution and add 0.2 g Coomassie blue. Stir to dissolve. Additional stain and/or destain solution may be necessary, depending on size of class and containers used.

6. If agarose slipping off the slide presents a problem, the melted gel can be placed in 35 mm plastic petri dishes. All other steps are carried out as described for the microscope slides.

Method

Part A: Preparation of Slides with Sample Wells

1. Each group will use three slides for the experiment. If desired, an additional slide can be prepared for practice sample loading. Clean all slides by placing a small amount of alcohol on each slide and wiping dry with paper towels. After cleaning, do not touch the clean surface with your fingers.

2. Label the three slides to be used for the experiment at one end with your group designation, and number them consecutively 1, 2, 3. Be careful not to touch the clean surface with your fingers.

3. Using a 5 ml pipet, carefully pipet 3.5 ml of melted agarose, allowed to cool slightly, onto each slide by moving the pipet tip along the center length of the slide and dispensing the agarose. The melted agarose should spread evenly over the surface of the slide, without running over the edges. Repeat with the remaining two slides.

Method, continued

NOTE: If you are not successful at covering the slide with agarose, wipe off the slide, clean the slide again with alcohol as described in step 1, relabel the slide, and try again.

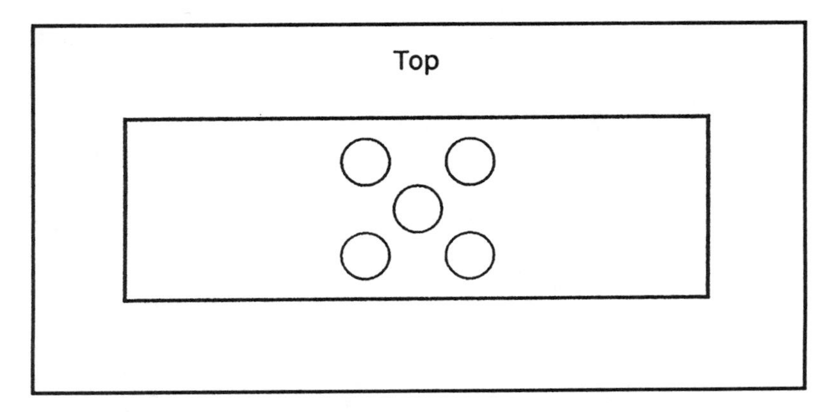

Figure 4. Template for cutting wells

4. Allow the agarose to solidify. This will take approximately 10 min., at which time the gel will appear slightly opaque.

5. For the three slides to be used for the experiment, place the template under one of the slides so that the pattern is in the center of the slide. The distances between the wells is important. Try to follow the template as accurately as possible.

6. Cut the five wells with the well cutter, using a gentle punching motion. Remove the agarose plugs by lifting the edges with a toothpick or spatula.

   NOTE: If well placement is not accurate, there should be enough room on the slide to recut the wells after repositioning the template under the slide.

7. Repeat steps 5 and 6 with the remaining two slides.

8. If a slide is to be used for practice loading, cut two rows of four wells each, using the well cutter. The positions of these wells are not critical, but leave enough space between wells so that the agarose plugs can be easily removed. Remove the agarose plugs as in step 6.

Method, continued

Part B: Loading the Samples

1. If practice loading is desired, load practice loading solution into the wells of the practice slide as follows:

   a. Squeeze the pipet bulb to slowly draw the liquid up into the pipet. The sample should remain in the lower portion of the pipet.

   b. While keeping the pipet tip above the tube containing the practice solution, slowly squeeze the pipet bulb until the sample is nearly at the opening of the pipet tip.

   c. Place the pipet tip just over, not inside, the sample well. Maintain steady pressure on the pipet bulb to prevent sample from being drawn back up into the pipet.

   d. Slowly squeeze the pipet bulb to eject three drops of sample. The well should appear full. Be careful not to overfill the wells and cause spillage on the agarose surface. Such spillage would affect your results.

2. Fill the center wells of all three slides with 3 drops (approximately 30 microliters) of antiserum (Ab, tube A), using a transfer pipet. Follow the procedure described for practice loading in step 1. Use the same pipet to fill all three center wells.

3. Fill the outer wells with 3 drops of antigen (tubes B, C, and D), using a clean transfer pipet for each antigen, as follows:

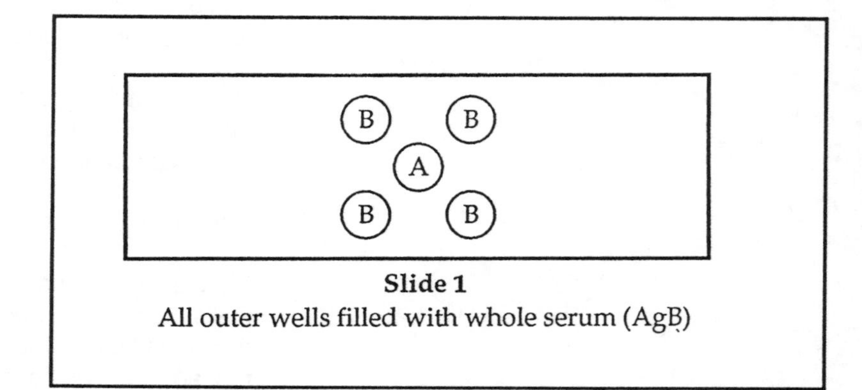

Slide 1
All outer wells filled with whole serum (AgB)

Figure 5

Method, continued

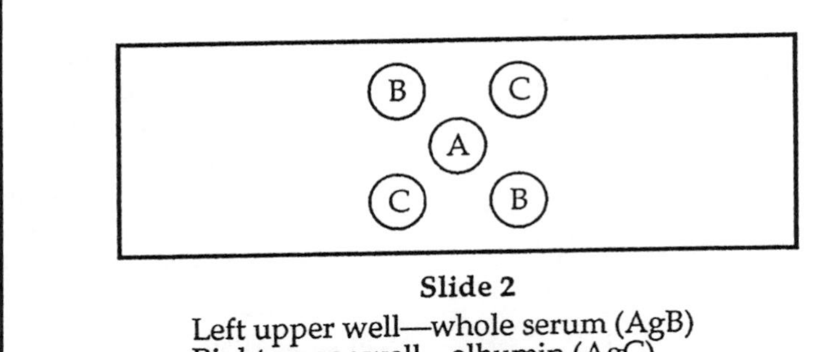

Slide 2
Left upper well—whole serum (AgB)
Right upper well—albumin (AgC)
Left lower well—albumin (AgC)
Right lower well—whole serum (AgB)

Figure 6

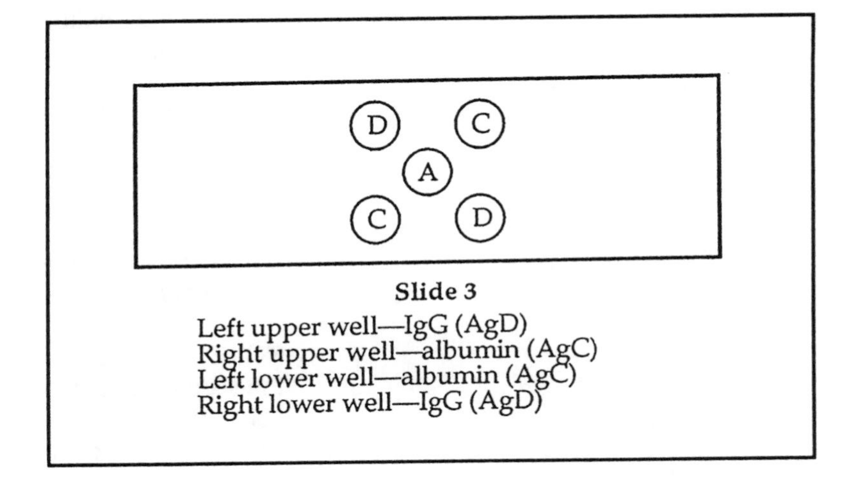

Slide 3
Left upper well—IgG (AgD)
Right upper well—albumin (AgC)
Left lower well—albumin (AgC)
Right lower well—IgG (AgD)

Figure 7

4. Carefully place the slides in the incubation chamber on top of the wet paper towel layer. Cover the chamber with plastic wrap and let incubate at room temperature overnight to allow precipitin lines to form. The precipitin lines should be visible in 24 hours.

5. Predict the patterns that will be formed by the precipitin lines, based on the relationships between the antigens and the information given in the introduction. Sketch your predictions on the diagrams of the three slides above.

Method, continued

Part C (optional): Staining slides

NOTE: Precipitin lines often can be seen more clearly if they are stained with Coomassie blue, a dye which binds to most proteins. This procedure is optional, since it requires 1-2 days total time, and the instructor may elect to omit the staining or to perform the staining his/herself. Coomassie blue staining requires that unprecipitated proteins be removed prior to staining, or the entire slide will be blue. Soaking the slides in several changes of Tris buffer will allow unprecipitated proteins to diffuse out of the gel, while precipitated proteins will remain in the gel (this is referred to as a dialysis procedure, although no semipermeable membrane is involved). For the staining procedure:

1. Cover slides with Tris buffer, 0.01 M pH 7.0, and incubate at room temperature for at least 4 hours, changing the Tris buffer once.

2. Remove buffer and cover the slides with 0.1% Coomassie blue stain and let sit 30-60 min.

3. Remove stain and cover slides with destain. Change destain when it becomes blue (about one hour), add fresh destain, and repeat until gel is clear enough to visualize blue precipitin lines.

4. Remove destain, cover dish and store slides in the refrigerator until results are observed.

NOTE: The agarose has a tendency to come off of the slides when the slides are submerged in liquid. Add liquids very gently. When solutions are changed, it may be easiest to carefully transfer the slides to a new dish containing the new liquid.

Results

1. Carefully hold a slide up so that the overhead room lights shine through it (or place the slide on a light box if available). The precipitin lines will be visible as opaque white arcs on each slide, where antibody and antigen complexes precipitated.

2. Draw the patterns of the precipitin lines observed on each slide. The precipitin lines should be seen easily. The expected results are shown in Figure 8. However, the intersections of the lines are often fainter and more difficult to see, making interpretation difficult. The Coomassie blue staining is recommended to increase visibility of the lines. A complete lack of precipitin lines may be due to:

   a. disproportionate pipetting between wells
   b. inconsistent distances between wells
   c. inactive antiserum
   d. error in preparing antigen sample, e.g., wrong concentration

3. Compare the patterns obtained to those illustrated in Figure 3. For each slide, determine relationship between the antigens which is indicated by the results. Do the results indicate identity, partial identity, or non-identity between the antigens?

4. Are the relationships indicated by the precipitin line patterns consistent with what you predicted, based on the contents of each antigen sample? Explain.

Hints/Troubleshooting, continued

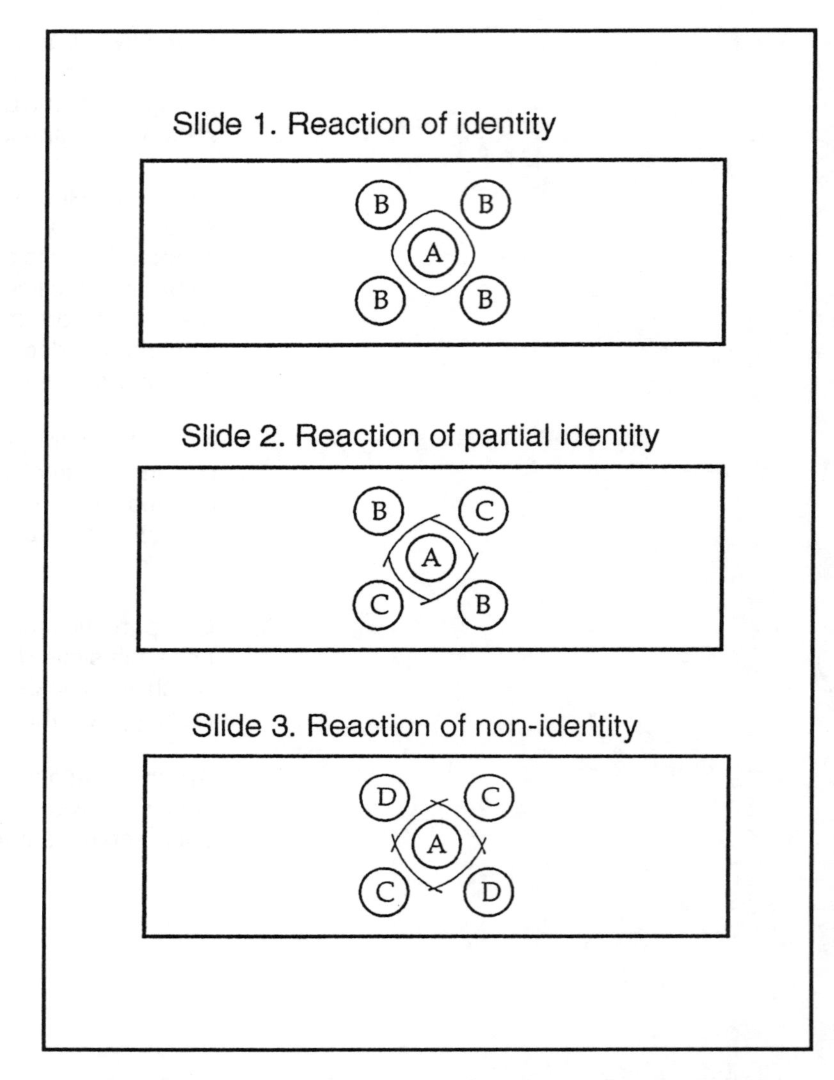

Figure 8. Predicted results for double diffusion experiment

UNIT 5: ANTIBODY SPECIFICITY AND DIVERSITY

MODULE 15: IMMUNOELECTROPHORESIS OF RABBIT SERUM PROTEINS

Baldwin King
Edward C. Kisailus

* Introduction
* Safety Guidelines
* Experimental Outline
* Materials
* Pre-lab Preparation
* Method
* Results

Introduction

Immunoelectrophoresis is a method used in both clinical and research laboratories for separating and identifying proteins on the basis of their electrophoretic behavior and their immunological properties. The introduction of these proteins such as rabbit serum proteins (antigens or immunogens) into another animal such as a goat elicits the production in the host (goat) of other proteins known as antibodies (immunoglobulins). The interaction between the antigen and its antibody is both strong and highly specific. If solutions of antigen and antibody are mixed in different ratios, it is found that at a specific ratio known as the equivalence point, the binding is maximized and a precipitate of antigen-antibody complex is formed. In the clinical laboratory, this technique is used diagnostically. It is utilized in examining certain serum abnormalities especially those involving immunoglobulins, and also semi-quantitative protein analysis of urine, cerebrospinal fluid, pleural fluids, etc. In the research laboratory, this technique may be used to monitor antigen and/or antibody purifications, to detect impurities, to analyze soluble antigens of plant and animal tissues and microbial extracts.

In immunoelectrophoresis, the proteins are first subjected to an electric field and separated by zone electrophoresis on the basis of their different charge-to-mass ratios. The electric field is then shut off and antibodies to the proteins are introduced into the system. Diffusion of both antigen and antibody takes place and at a particular locus, the equivalence point is reached resulting in precipitation. A number of different configurations have been developed to separate, characterize and quantify proteins depending on the emphasis. In classical immunoelectrophoresis, the antibody is added in a trough cut out of the electrophoretic gel after electrophoresis.

This module is to demonstrate the use of immunoelectrophoresis to separate and characterize a mixture of (rabbit serum) proteins as well as to examine the specificity of the antigen-antibody interaction.

Safety Guidelines

Boiling agarose can spatter and cause severe burns. When heating agarose wear safety goggles and hot gloves. Latex gloves should be worn throughout the procedure, especially when handling solutions or gels containing methylene blue. Prior to turning electrophoresis power pack on, be sure that the chamber is level and that the work surface is dry. Wear gloves and safety goggles when operating electrophoresis apparatus.

Experimental Outline

Three hour laboratory with overnight incubation.

Prepare and pour gel - 20 minutes
Load samples - 10 minutes
Electrophoresis - up to 2 hours
Load samples - 10 minutes
Data analysis - 30 minutes

Prepare and pour agarose gel

Transfer cooled gel to electrophoresis chamber

Cut 2 small holes in gel

Fill chamber with buffer

Load rabbit serum solution

Electrophorese

Cut 2 parallel troughs in gel

Load troughs with anti-rabbit IgG and anti-rabbit whole serum

Allow precipitates to develop

Materials

2 g of agarose
500 ml of TAE (Tris-Acetate-EDTA) buffer, pH 8.0
10 μl of rabbit whole serum
200 μl of goat anti-rabbit serum
10 μl of rabbit albumin
200 μl of goat anti-rabbit IgG
Horizontal electrophoresis apparatus

A kit, which contains all the critical reagents required for this experiment, is available from EDVOTEK (1-800-EDVOTEK).

Pre-lab Preparation

TAE (50X stock, 1 liter)

242 g Tris base
57.1 ml glacial acetic acid
37.2 g $Na_2EDTA \cdot H_2O$
pH 8.5

Dilute antiserum in TAE if necessary

Method

1. Prepare buffer as follows:

   Stock solution- Dissolve 121 g trishydroxymethyl-amino-methane (Tris) and 17 g of ethylenediaminetetracetate, disodium salt (EDTA) in 450 ml of deionized water. Adjust pH to 8.0 with glacial acetic acid (25 ml). Make up to 500 ml with water.

   Working buffer - dilute stock 50 times

2. Prepare agarose gel as follows:
   Mix 0.5 g of agarose and 50 ml of diluted TAE buffer in a 250 ml Erlenmeyer flask. Dissolve by boiling in a microwave oven or on a hot plate. Allow to cool to 50 degrees Celsius. If gel does not stay on slide and runs off edges, clean slides in alcohol. Temperature of molten gel is critical to be at 50 degrees Celsius.

3. Coat a glass plate by applying 2 % agar solution with a small brush.

4. Cast the gel on to the glass plate by pouring the molten gel to a thickness of 1-2 mm and allow to cool.

5. Cut 2 or more small holes in the gel by punching the agarose gel with a pasteur pipette connected to a water aspirator.

6. Transfer the gel to the electrophoresis apparatus and add enough buffer to just cover the gel.

7. Add 2 drops of a bromophenol blue solution to the rabbit albumin and the rabbit whole serum solutions.

8. Load rabbit albumin into left well and rabbit whole serum into the right well with a micropipette or capillary tubing.

Method, continued

9. Elevate slide 2-3 mm above electrophoresis buffer solution. Make contact between the buffer and agarose by using a wick of filter paper presoaked in electrophoresis buffer. One end of the wick should overlap the end of the gel by 3-4 mm and the other end submerged in electrophoresis buffer.

10. Electrophorese at 50 V and 30 mA for 1 hour

11. Carefully remove the gel from the electrophoresis chamber. Cut two troughs, one between the two wells and the other to the right of the whole serum well. Use a sharp blade to delineate the troughs and a Pasteur pipette tip to remove the cut gel.

12. Load the left trough with goat anti-rabbit whole serum and the right trough with goat anti-rabbit IgG.

13. Place the gel in a closed humidifying chamber containing moistened paper towels and allow diffusion to take place over a 24 hour period or until a visible precipitate is formed.

Results

1. Note the formation of arcs of precipitate in the gel.

2. Identify the number of proteins in the rabbit whole serum from the number of arcs of precipitate.

3. Identify albumin and IgG in the whole rabbit serum from comparison with the pure albumin and the pure anti-IgG segments.

The actual number of arcs of precipitate will depend on the source of your rabbit serum and the efficiency of separation. However the following general pattern should be seen:

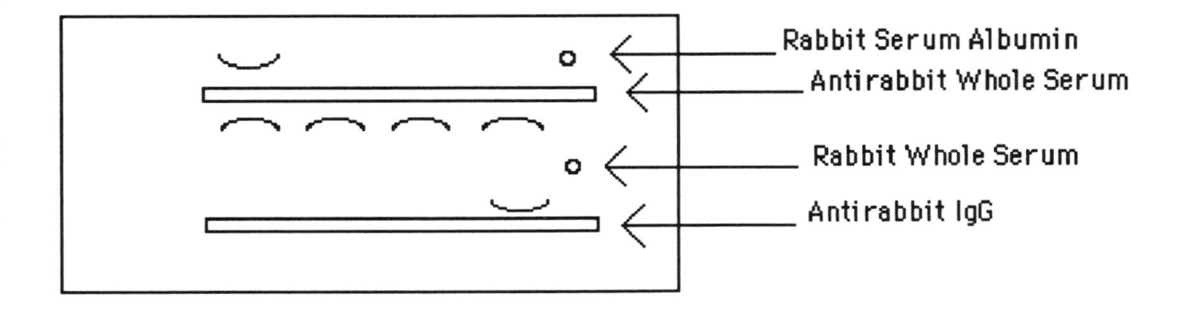

UNIT 5: ANTIBODY SPECIFICITY AND DIVERSITY

MODULE 16: SPECIFIC AGGLUTINATION AND AGGLUTINATION INHIBITION OF YEAST CELLS BY LECTIN

Edward C. Kisailus

* Introduction
* Safety Guidelines
* Experimental Outline
* Materials
* Pre-lab Preparation
* Method
* Results

Introduction

The phenomenon of antibody binding to insolubilized antigen as part of a cell membrane or cell wall may lead to cross-linking of the cells called agglutination. Agglutination occurs as long as there is an adequate concentration of antigen on the cell surface and the cross-linking protein is at least divalent. The affinity constant is the quantitative measure of the strength of binding of the agglutinating protein. It must be great enough to maintain an equilibrium in order for two or more cells to be bound together by the antibody.

Antibody agglutination of cells is a major way in which the immune system functions to clear the body of the antigen-containing cell. Agglutination in vitro is a valuable diagnostic and research tool to give one a first approximation of the presence or absence of an antigen. Besides antibody there is a class of naturally occurring carbohydrate-binding proteins which will agglutinate cells. These proteins, typically found in plant seeds and animal tissues, are called lectins. They are defined as proteins of non-immunologic origin, having a carbohydrate-binding specificity, and lacking an apparent enzyme activity.

The use of a lectin as an agglutinating protein exhibits their function in the living organism. In leguminous plants this binding may be important to capture microorganisms that will be enclosed in nodules. Other plant lectins may bind soil microorganisms that are potential pathogens - a rudimentary plant immune system. Concanavalin A, a lectin isolated from Jack bean meal, has a carbohydrate-binding specificity for alpha-linked mannose and glucose residues. The specificity of the lectin combining site is determined by observing the degree of agglutination inhibition by using a bank of mono- and disaccharide inhibitors. Antibody combining site specificity can be determined using a similar inhibition method. The inhibiting compounds for carbohydrate-specific antibody are generally more complex in size, makeup, and conformation due to the larger antibody combining site relative to that of most lectin combining sites.

Yeast cells, *Saccharomyces cerevisiae*, have a cell wall component called mannan that is a polymer of alpha-linked mannose residues. Concanavalin A will bind to mannan and with an adequate concentration of cells agglutinate the yeast cells. There is no biological importance to this phenomenon. However, it does exhibit agglutination quite well.

Introduction, continued

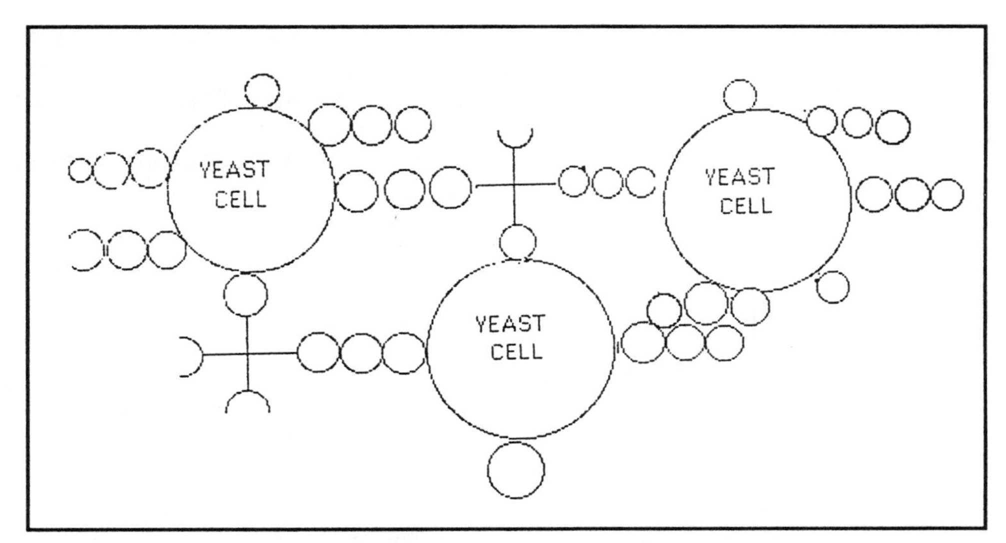

Figure 1. To Illustrate Agglutination

Antibodies and lectins are proteins with combining sites of defined specificity. The nature of the combining site specificity is determined by studying the size, shape, and chemical groupings within the combining site. The interaction of chemical groupings of a ligand and a combining site is required to maintain a "fit" or association which stabilizes the complex. Antibodies especially to protein and polysaccharide antigens have a combining site specificity which when probed using polypeptides (short chains of amino acids) or oligosaccharides (short chains of carbohydrates) conforms to a site of variable depth and conformation. Molecules with the proper shape and chemical grouping to fill the combining site will inhibit the binding of other molecules with a like specificity for the combining site. Lectins do not have combining sites as large and complex as carbohydrate-binding antibodies.

An approximation of antibody and lectin combining site specificity can be made by studying a battery of molecules of known chemical makeup and structure. Lectin combining site specificity can be investigated using agglutination inhibition. The inhibiting molecules are soluble. The molecule to which the lectin is being inhibited from binding is that which is on the cell surface. The competition for the combining site between the insolubilized cell surface component and the solubilized inhibitor is concentration dependent and also dependent on complementarity of fit as described above. The more complementary an inhibitor is to the combining site the more effectively it will compete in binding to the combining site. Complementarity includes, in part, the chemical interaction that occurs between an inhibitor and a combining site.

Introduction, continued

Two molecules are competing for the combining site in agglutination inhibition. The insolubilized molecule on the yeast cell is called mannan, a yeast cell wall protein-polysaccharide complex where the polysaccharide is mainly alpha-linked mannose chains; and its inhibitor is a soluble inhibitor molecule of a mono or disaccharide nature.

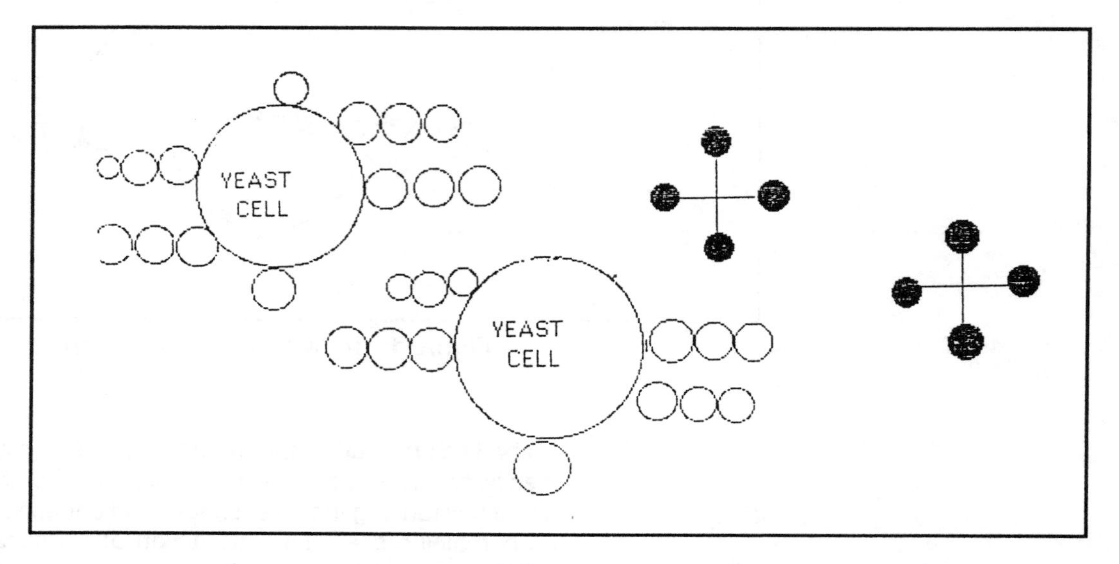

Figure 2. Agglutination Inhibition

Safety Guidelines

Safety - cleanup, waste disposal

Yeast - keep away from cell culture area to prevent contamination. Autoclave all tubes containing yeast

Con A solutions can be disposed of in sink drain

PBS can be disposed in sink drain.

Experimental Outline

Timetable of Events

Pipetting - 20 minutes
Ice incubation - 10 minutes
Macroscopic observation - 5 minutes
Wet mounts and microscopy - 5 minutes
Counting Cells/agglutinate - 40 minutes
Histogram - 20 minutes
Data Analysis - 20 minutes

Procedures

1. Label test tubes
2. Pipet control and lectin solutions
3. Pipet inhibitor solutions
4. Pipet yeast suspension into control and lectin solutions
5. Ice bath for ten minutes
6. Observe for agglutination macroscopically and microscopically

Materials

Yeast cell suspension (Fleischmann's dry yeast)
Phosphate buffered saline (PBS)
Concanavalin A, Sigma Chemical and dilutions (in PBS) of 1 mg ml, and 0.1 mg/ml
.1M Lactose, 0.5M D-Galactose, 0.5M D-Glucose, 0.5M alpha-methyl D-mannoside, 0.5M Sucrose
Pasteur pipettes
1, 5, and 10 milliliter pipettes
15 mm diameter test tubes
Microscope slides, cover slips
Microscope
Hand lens

A kit, which contains all the critical reagents required for this experiment, is available from EDVOTEK (1-800-EDVOTEK).

Pre-lab Preparation

PBS: PBS can be purchased as a 1x or 10x solution, or as a powder. If prepared in the laboratory, dissolve the following in 150 ml distilled water:

2.19 g NaCl
1.28 g KH_2PO_4
2.63 g $Na_2HPO_4 \cdot 7H_2O$

Adjust pH to 7.2, add distilled water to bring volume to 250 ml

Yeast suspension - [Resuspend 600 mg of Fleishmann's dry yeast in 50 ml PBS (need 10 ml suspension for whole module)] - 10 minutes (Can be stored in refrigerator for 2 days)

Con A dilution from lyophilized powder - 20 minutes (can be stored in refrigerator for days if solution is kept sterile or for weeks if stored frozen at -20°C)

Con A dilution from concentrated solution - 10 minutes (can be stored in refrigerator for days if solution is kept sterile or for weeks if stored frozen at -20°C)

Method

Agglutination

1. Label four 15 mm tubes:
   Tube A: lectin 500
   Tube B: lectin 100
   Tube C: lectin 10
   Tube D: PBS control

2. Using a 1 ml pipette, pipet 0.5 ml of 1 mg/ml Con A into Tube A.

3. Using a 1 ml pipette, pipet 0.1 ml of 1 mg/ml Con A, and 0.4 ml of PBS into Tube B. Mix.

4. Using a 1 ml pipette, pipet 0.1 ml of 0.1 mg/ml Con A, and 0.4 ml of PBS into Tube C. Mix.

5. Using a 1 ml pipette, pipet 0.5 ml of PBS into Tube D.

6. Using a 5 ml pipette, pipet 0.5 ml of stock culture of yeast cells into each tube - tubes A, B, C, and D. Mix.

7. Place the tubes on ice for 10 minutes, mix occasionally - every three or four minutes.

8. Mix and observe for agglutination.

9. Mix and prepare a wet mount by placing one drop of the suspension on a microscope slide and covering the drop with a 22 x 22 mm coverslip.

10. Place the slide and wet mount on the stage of a light microscope. Find the field of cells beneath the cover slip by using the 10X objective. Then view the cells by light microscopy at 400x magnification.

Agglutination Inhibition

1. Label seven 15 x 150 mm tubes:
   Tube A: lectin 500
   Tube B: Lactose
   Tube C: Galactose
   Tube D: Glucose
   Tube E: alpha-methyl mannoside
   Tube F: Sucrose
   Tube G: PBS control

2. Pipet 0.5 ml of 2 mg/ml Con A into each tube <u>except</u> Tube G.

Method, continued

3. Pipet 0.5 ml PBS into Tube A, and 1.0 ml PBS into Tube G.
4. Pipet 0.5 ml carbohydrate inhibitor into Tubes B, C, D, E, and F. Mix.
5. Pipette 1.0 ml of stock culture of yeast cells into each tube. Mix.
6. Place the tubes onto ice for 10 minutes, mix occasionally - every three or four minutes.
7. Mix and observe for agglutination
8. Mix the cells, prepare a wet mount and view cells by light microscopy at 100x magnification.

Results

For observing agglutination macroscopically use a hand lens to observe the suspension of yeast cells. Score each tube using - to ++++ system with the PBS control (Tube D) being the negative (-) score control.

For microscopic quantitative analysis randomly select fields of cells on the microscope slide. The objective is to view random fields of cells and count the single cells and number of cells in agglutinates. So as not to bias results by looking for a "good field with lots of cells" randomize the selection of fields. Random selection is best accomplished by moving the microscope stage vertically or horizontally without viewing the movement through the ocular(s). Move the stage slightly for high powered fields since a small move changes the field completely. The goal is to gather a sufficient quantity of data to graph results so that a conclusion(s) can be made. Ideally every cell and/or agglutinate on the slide should be scored. This is impractical. Arbitrarily choose to count cells and agglutinates in ten high powered fields (hpf). Single cells are counted as one, agglutinates are categorized in the following way 2-3 cells/agglutinate, 5-9 cells/agglutinate, 10-14 cells/agglutinate, 15-20 cells/agglutinate, and 20+ cells/agglutinate. The numbers of cells in an agglutinate can be counted by focusing in and out of the agglutinate using your fine focus adjustment. Cells above and below the plane of focus will come into view and can be scored or counted. If an agglutinate obviously has more than 20 cells don't bother to count the cells rather score it as a 20+.

Results, continued

Control Tube

Scan the Control tube D wet mount using the 4X or 10X objective for a field with a significant number of cells. Adjust the microscope to a higher power of magnification, preferably 400X and observe the cells. Count the number of free cells and cells in agglutinates in this first high powered field (hpf). Move the stage horizontally and vertically and after each move count the number of free cells and cells in agglutinates. Count 10 hpf. Make a table indicating the number of cells in a clump and the number of free cells. This Control tube data is your background nonspecific agglutination data.

	Number of cells					
hpf	1	2-3	5-9	10-14	15-20	20+
1						
2						
3						
4						
5						
6						
7						
8						
9						
10						

Experimental Tube

Prepare a wet mount of experimental tubes A - C. Agglutinated cells will appear as clumps of cells in a "cluster of grapes" appearance. Scan for a field with a significant number of cells. Adjust the microscope to a higher power setting, preferably 400X and observe the cells. Arbitrarily pick a field with a significant number of cells. Count the number of free cells and cells in agglutinates in each tube. Count agglutinates and/or free cells in 10 hpf. Complete the table by indicating the number of cells in a clump and the number of free cells.

UNIT 5: ANTIBODY SPECIFICITY AND DIVERSITY

MODULE 17: GENERATION OF DIVERSITY IN LYMPHOCYTE RECEPTORS

Ah-Kau-Ng

* Introduction
* Safety Guidelines
* Experimental Outline
* Materials
* Pre-lab Preparation
* Method
* Results

Introduction

B and T lymphocytes are two types of leukocytes (i.e. white cells) that are responsible for antigen specific immunity. B cells express antibody molecules on their surface as antigen receptors, and will develop into antibody secreting plasma cells upon activation by their corresponding antigens. T cells also carry surface receptors for antigens, termed simply T cell receptors or TcR. They include helper T cells, which 'help' B cells make antibodies, and cytolytic T cells which kill target cells specifically.

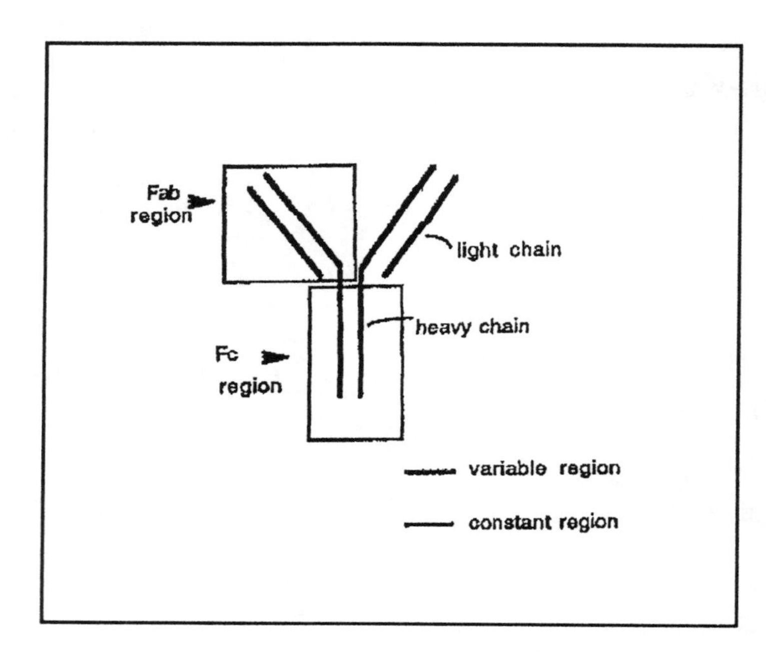

Figure 1

An antibody (i.e. immunoglobulin or Ig) is a protein molecule with two identical heavy (H) chains and two identical light (L) chains (see Figure 1). Both the H and L chains consist of a variable (V) region that constitutes the antigen combining site, and a constant (C) region that determines the isotype (e.g. IgG, IgM, IgA, IgD and IgE for H chains; κ, λ for L chains) of an antibody. T cell receptors (TcR) have a heterodimeric protein structure which exists either as a α/β or a γ/δ form. Each (TcR) polypeptide subunit (i.e., α, β, γ, δ) also carries an antigen-recognizing V region, and a C region of unknown function.

It is generally assumed that a human body (other mammalian species as well) is able to produce enough antibodies or TcR, each with a different specificity, to cope with all foreign antigens that may ever exist (possibly in the range of 10^8). This assumption, however, does not agree with the estimate that the human genome has only approximately 10^6 genes in reserve. How, then, can our immune system possibly generate such a diversity that far exceeds our total genetic capability? An answer to this longtime puzzle has become clear only recently, due largely to the molecular biology studies pioneered by Susumu Tonegawa and others. It was discovered that a considerable diversity of T and B cell specificities could be generated by recombining scattered, discrete segments (i.e. DNA sequences that represent "minigenes") of Ig genes or TcR genes. The gene rearrangement events occur in precursor cells during their development into mature T or B cells. Once completely rearranged, the DNA sequence (i.e. specificity) of a full-length Ig gene (of a mature B cell) or a TcR gene (of a mature T cell) is permanently fixed.

Take for example the generation of a complete gene that will code for a full length heavy (H) chain of an antibody. A complete H gene consists of a variable (V) region gene segment, a diversity (D) gene segment, a joining (J) gene segment and a constant (C) region

Introduction, continued

gene segment. The juxtaposed (V-D-J) gene segments jointly code for the variable region of the H chain, which determines antibody specificity; the C gene codes for the constant region which does not contribute to antibody specificity. Before H chain gene rearrangement there are by estimate approximately 250 V, 10 D and 5 J gene segments in each precursor cell of a B lymphocyte. During development into a B cell one V, one D and one J gene segment will combine in a precursor cell, and there are potentially 12,500 (i.e. 250 x 10 x 5) ways a combination could result! (It works in a way like the number-locks.) Similarly, by combining at random the V and J gene segments (approximately 200 V and 10 J) of the light chain (L) genes, B cells can generate 2,000 L chain combinations. Pairing of H and L chains (H x L) could thus provide up to more than 10^7 specificities. Additional diversity is created by imprecise joining of the various gene segments, and by point mutation, altogether leading to billions of antibody possibilities. Similarly, during T cell development TcR gene segments in precursor cells undergo rearrangement, resulting in the generation of enormous diversity in T cell specificities.

In this module, you will perform a 'dry lab' to investigate the gene rearrangement mechanism by which the immune system generates diversity in antigen recognition by B cells (i.e. using antibody) and T cells (i.e. using TcR). Paper clips with different colors will be used to represent different minigenes or DNA segments of Ig and TcR genes. You will be instructed to combine these paper clip genes to construct a series of complete IgH genes and TcR-β genes that code respectively for antibodies and T cell receptors of different specificities.

Safety Guidelines

None, except avoid sticking your fingers with the sharp ends of paper clips.

Experimental Outline

Each of the two exercises could be completed within an hour. It is recommended that Exercise A, and Exercise B be performed in sequence.

Sort out colored paper clips

Perform Experiment A: Ig gene rearrangement

Perform Experiment B: TcR gene rearrangement

Group paper clips according to color and return to instructor

Materials

Paper Clips: red (R), pink (Pi), orange (O), yellow (Y), blue (Bl), green (G), purple (Pu), white (W), black (B), silver (S).
Each paper clip represents a "mini-gene" or gene segment.

(Alternatively, paper clips tagged with color adhesive tapes, or strings or beads with corresponding colors can be used.)

Pre-lab Preparations

No special pre-lab preparations are required.

Method

Experiment A

Immunoglobulin (Ig) gene rearrangement

1. Obtain paper clips with different colors from instructor.

2. Use the following letter keys for the paper clips.

Black - B	Blue - Bl	Green - G
Orange - O	Pink - Pi	Purple - Pu
Red - R	Silver - S	White - W
Yellow - Y		

Method, continued

3. Assume that each paper clip represents a gene segment, use your paper clips to represent discrete DNA segments of Ig heavy chain (H) genes as follows:

   Variable region (V) genes: V_1 (Bl), V_2 (G), V_3 (O), V_4 (Pi)
   Diversity region (D) genes: D_1 (Pu), D_2 (R)
   Joining region (J) genes: J_1 (W), J_2 (Y)
   Constant region (C) gene: C (B)

   Use the silver (s) paper clips to represent intervening sequence (i.e. introns).

4. Arrange your paper clips as shown in Figure 2 to construct a hypothetical embryonic or germ line DNA sequence of Ig gene before rearrangement.

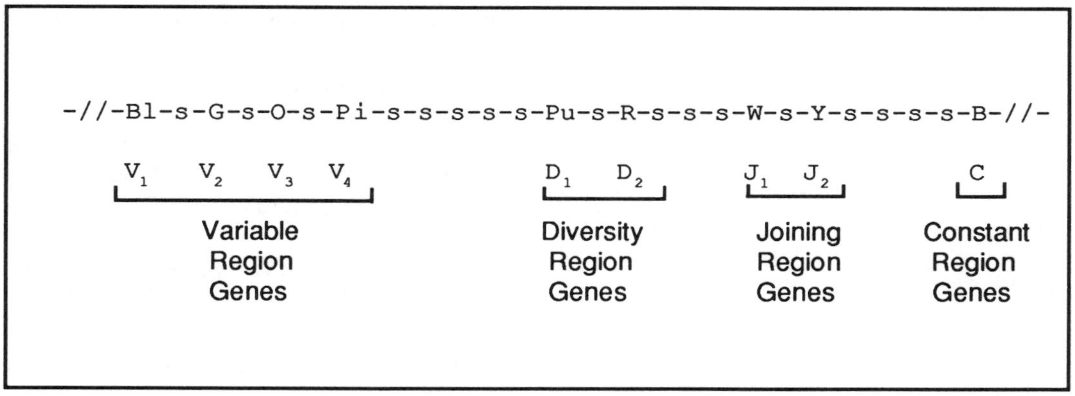

```
-//-Bl-s-G-s-O-s-Pi-s-s-s-s-s-Pu-s-R-s-s-s-W-s-Y-s-s-s-s-B-//-
```

V_1 V_2 V_3 V_4	D_1 D_2	J_1 J_2	C
Variable Region Genes	Diversity Region Genes	Joining Region Genes	Constant Region Genes

Figure 2. Germ-line or embryonic configuration or Ig heavy chain gene segments before rearrangement.

5. Each complete heavy chain gene consists of one V gene, one D gene, one J gene and one C gene. Using your fingers as recombinase enzymes, construct a complete heavy chain gene by combining one of the four V genes, one of the D genes, one of the J genes and the C gene (for simplicity, only one C gene is considered in this exercise). You can assume that any V gene can combine with any D gene, which in turn can combine with any J gene. Record the recombinant gene sequence (e.g., V_1-D_2-J_3-C). Return the paper clips to the original DNA sequence as shown in Figure 3.

6. Again, follow the instruction in (5) and construct another complete heavy chain gene sequence with a configuration different from the previous one. Record the new sequence. Return the paper clips to the original DNA strand.

Method, continued

7. Assume that you have been infected by five different infectious agents and your immune system responded by producing a different Ig heavy (H) chain specific for each invader. Demonstrate how this could be achieved by combining the 'paper clip' genes in Figure 1 to construct five complete H genes that will code for 5 distinct heavy chains (representing five specificities). Record your five combinations as in (4).

8. Disconnect your paper clip chains and group the paper clips by color.

9. Proceed to EXPERIMENT B.

Experiment B

T cell receptor (TcR) gene rearrangement

1. As in EXPERIMENT A

2. Assume that each paper clip represents a gene segment, use your paper clips to represent discrete DNA segments of TcR β chain genes as follows:

   Variable regions (V) genes: V_1(B), V_2(W), V_3(Y), V_4(Pu)
   Diversity region (D) genes: D_1(R), D_2(G)
   Joining region (J) genes: J_1(O), J_2(Pi)
   Constant region (C) genes: C(Bl)

   Intervening sequence: Intron (s)

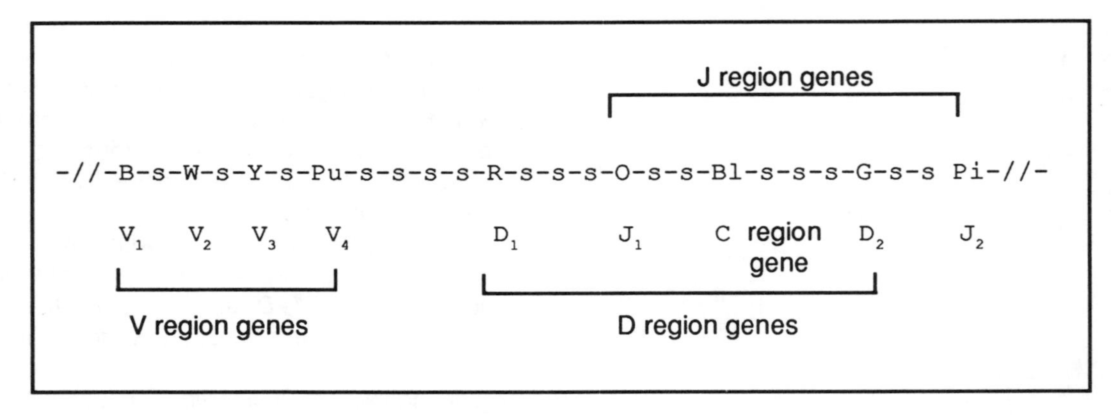

Figure 3. Germ-line or embryonic configuration of TcR β chain gene segments before rearrangement. Paper-clip TcR β gene sequence.

Method, continued

3. As shown in Figure 3, arrange your paper clips to assume a hypothetical germ-line configuration of TcR β gene sequence before rearrangement.

4. As assumed in Experiment A, you have been infected by five different infectious agents and your B cells responded by producing an antibody specific for each organism. Again, assume that for antibody production each antibody producing B cell must collaborate with a T helper cell carrying a surface receptor (i.e., TcR) specific for the corresponding organism. Using the paper clip gene sequence shown in Figure 4, construct five complete genes that could code for five distinct TcR β chains. As with the Ig heavy chain, the complete gene for a TcR β chain includes one V, one D, one J and one C region gene. Record the five combinations you constructed.

Results

1. Using a format as shown in the example below, illustrate in parallel the DNA sequences of all the five complete Ig heavy chain genes you constructed in step 4 of the Experiment A. Try using colors that correspond to the gene segments to accentuate the homologous and non-homologous regions among the five different DNA sequences.

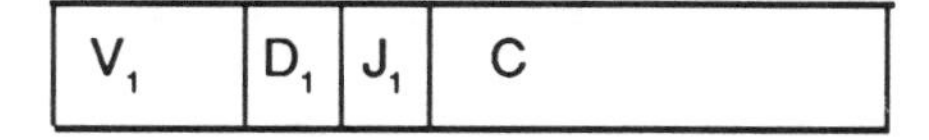

2. As instructed above, illustrate in parallel the DNA sequences of all five complete TcR β chain genes you have constructed in step 4 of the Experiment B.

3. Do you think your results demonstrate gene rearrangement as an effective mechanism to generate diverse gene sequences? Discuss your opinion(s).

4. As a conclusion of your experiments, write a paragraph to compare the strategies employed by T and B cells to generate diversity in their receptors for antigens.

UNIT 5: ANTIBODY SPECIFICITY AND DIVERSITY

MODULE 18: ENZYME-LINKED IMMUNOSORBENT ASSAY (ELISA)

Ah-Kau Ng

* Introduction
* Safety Guidelines
* Experimental Outline
* Materials
* Pre-lab Preparation
* Method
* Results

Introduction

The body's immune system plays an important role in host defense mechanisms against foreign invaders. One major response of the immune system to infection is the production of antibodies, (i.e., immunoglobulins or Ig) in response to the stimulation of antigen(s) of the infecting agent. The antibody formed reacts specifically to the antigen, leading to the eradication of the particular infecting agent and no other. The antibody proteins circulate in the blood stream and can be detected in the serum of the infected individual.

Many laboratory procedures (i.e., assays) have been developed to measure antibodies and to detect the specific interaction between antibodies and antigens. These immunoassays include immunoprecipitation test (e.g. gel diffusion), agglutination test, complement fixations test, immunofluorescence test, radioimmunoassay (RIA) and enzyme-linked immunosorbent assay (ELISA), Western Immunoblot assay, etc. Using these immunoassays the antibody response of a host against infection can be followed, and the results obtained can contribute to the diagnosis of the infection. For example, detection of serum antibody to HIV (AIDS virus) is considered a reliable criterion to identify infected individuals who have become seropositive.

Although initially developed for antibody measurement, immunoassays have also been adapted successfully to detect and quantitate antigen in unknown samples. For example, antibodies specific for HIV antigens can be produced by injecting the antigens into laboratory animals (e.g. rabbits), and the specific antibodies obtained can then be used to develop immunoassays for measuring HIV antigens in infected individuals. Due to their high specificity and sensitivity immunoassays are commonly used for a great variety of measurements of both antibodies and antigenic analytes (e.g. hormones, drugs, etc.) in research, analytical and diagnostic laboratories.

The basis of all immunoassays is an interaction between an antibody and a corresponding antigen. The interaction can in turn be conveniently detected by conjugating a measurable label to either the antibody or the antigen, such as a radioisotope, a fluorescent compound or an enzyme. An enzyme label is generally favored over either fluorescent compounds which require expensive equipment for detection, or radioisotopes which are a biohazard and unstable. Thus, in recent years, the enzyme-based immunoassay, generally referred to as enzyme linked immunosorbent assay (ELISA), has seen increasing popularity in a wide range of applications. Like the radioimmunoassays, there are two basic types of ELISA: (1) direct binding assays primarily for detection or quantitation of antibodies; and (2) competition enzyme immunoassays for the detection and quantitation of antigens.

Introduction, continued

In this module ELISAs (the direct binding assay type) will be used to (1) observe a specific interaction between an antigen (BSA) and its antibody (anti-BSA), and (2) determine by titration the amount (i.e. titer) of anti-BSA antibody in a serum preparation from a rabbit immunized with the BSA antigen. The whole experiment will require two 3-hour lab periods and will include the following steps. In the first lab period: (1) prepare antigen solutions and coat the antigens to a 96-well microtiter plate; and (2) prepare serial dilutions of rabbit serum solutions. In the next lab period: (1) add antibody solutions to antigens coated on the plate to allow antigen-antibody interaction; (2) add second antibody conjugated with an enzyme (e.g. horseradish peroxidase enzyme conjugated with goat antibodies specific for rabbit antibodies); this second antibody, referred to as anti-Ig, will react with the antigen-antibody complex that formed earlier; and (3) add a substrate solution that will react with the enzyme conjugate to form a color product. Theoretically, the intensity of color developed (i.e. absorbance or optical density) is proportional to the quantity of anti-BSA antibody in the antiserum sample. No significant color change will be observed if no anti-BSA antibodies are present in the rabbit serum. The overall ELISA procedures are outlined in Figure 1.

Safety Guidelines

Some of the chemicals are irritants and could cause harmful reactions. Before proceeding be sure to put on lab coat, safety goggles, and plastic gloves. Avoid skin/eye contact with any reagent and do not ingest. Flush spills and splashes with water for at least 15 minutes. Report any accident to the instructor. At the end of the experiment each day, dispose of waste by following laboratory guidelines and wash hands thoroughly before leaving the laboratory.

Standard lab safety guidelines for handling chemicals and biological samples are adequate here. Students must wear lab coat (or apron), gloves and goggles in performing the experiment. The substrate solution which contains TMB, a potential mutagen, and organic solvent should be handled with caution.

Experimental Outline

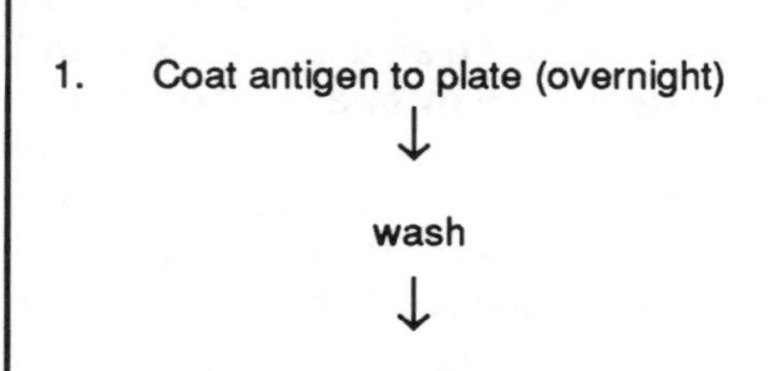

1. Coat antigen to plate (overnight)

↓

wash

↓

2. Add serum: any specific antibody attaches to antigen (1 hr)

↓

wash

↓

3. Add enzyme-labelled second antibody which attaches to first antibody (15 min at 37°C)

↓

wash

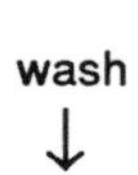

4. Add substrate (15 - 30 min)

↓

Amount hydrolysed =
color intensity =
amount antibody present

Figure 1.

Lab Period 1

(a) Pre-lab discussions of background information and protocol
(b) Coating antigens to microtiter plates
(c) Preparation of serial dilutions of rabbit sera

Lab Period 2

(a) Pre-lab discussion
(b) Saturation of antigen plate
(c) Addition of first antibody
(d) Addition of second antibody-enzyme conjugate
(e) Addition of substrate
(f) Read results

A flow chart of the ELISA protocol is shown in Figure 1.

Briefly, the four basic steps of the assay are:

1. Binding the Antigen - BSA or gelatin is added to separate wells of the microtiter plate. Allow time for the antigen to bind to the plate.

2. Antibody Addition - Plates are washed. Rabbit serum A containing antibody to BSA and rabbit serum B (normal rabbit serum without antibodies to BSA) are serially diluted and added to specific wells of the ELISA plate. This is incubated to allow for the antigen-antibody interaction.

3. Conjugate Addition- The plates are again washed. This will wash away antibody that has not reacted with the antigen on the plate. The conjugate (enzyme coupled to an anti-Ig or second antibody) is added. Conjugate A binds rabbit antibody. Conjugate B binds to mouse antibody. Both conjugates are linked to horseradish peroxidase in this case. Again the plates are incubated.

4. Addition of Substrate - Again, wash away unbound material. An enzyme substrate (a color forming chromogen in this case) is added. This reacts with the conjugate to produce a color indicating an antigen-antibody reaction.

Materials

PBS - Phosphate buffered saline (PBS) is provided as a 10X solution. Add 50 ml to a 500 ml graduated cylinder. Fill to the 500 ml mark with distilled H_2O for working solution. (Optional: add 0.5 ml of Tween 20 per liter of 1X PBS).

PBS (3% milk) - Blocking diluent is made by adding 3 grams nonfat dry milk to 100 ml 1X PBS.

Carbonate buffer 0.05M - sodium carbonate buffer, pH 9.6, containing 0.02% sodium azide.

Antigen I 1% Bovine Serum Albumin (BSA) - A vial is provided containing 0.1 grams of BSA. This must be mixed with 10 ml of carbonate buffer for working solution Antigen I.

Antigen II 1% Gelatin - Prepare a 1% solution by dissolving 0.1 grams into 10 ml of carbonate buffer. This must be heated until the gelatin is dissolved. Cool before using.

Rabbit Serum A - Contains antibodies against BSA. Provided in vial is 0.1 ml of a 1:10 dilution. Dilute this with 9.9 ml of PBS (3% milk) as follows. Empty contents of vial into a 15 ml test tube. Measure out 9.9 ml of PBS (3% milk). Use a small amount to rinse out vial. Empty this and remaining diluent into the 15 ml tube. Mix. Label tube #1 Rabbit Serum A 1:1000.

Rabbit Serum B Normal Rabbit Serum - Does not contain antibodies against BSA. Prepare as above for Rabbit serum A. Label #8 Rabbit Serum B 1:1000.

Anti-Rabbit Ig Conjugate - 0.2 ml provided. Keep frozen until ready to use. Dilute in 19.8 mls PBS (3% milk) in 50 ml tube. Label Rabbit Conjugate.

Anti-Mouse Ig Conjugate - Horseradish peroxidase conjugated antibody specific for mouse Ig. Store and dilute as for anti Rabbit Ig conjugate. Label Mouse Conjugate.

TMB Substrate - This is provided as solution A and B. Do not prepare until just before use. Mix 30 mls solution A with 30 mls of solution B. Label TMB. Keep from light.

Spectrophotometer, with microcuvette for 250 µl, ELISA Microplate Reader

96 well Microtiter plate

Pipetman, P 200

Multichannel pipet (optional)

Disposable pipet tips

Materials, continued

Water, deionized or distilled
Bovine Serum Albumin (BSA) Solution, Antigen I
Gelatin Solution, Antigen II
Phosphate buffer saline (PBS), 10X
PBS 1X with 3% milk, blocking solution and serum diluent
Carbonate buffer, antigen coating solution
Rabbit anti-BSA antiserum, rabbit serum A, 1:10 dilution
Anti-rabbit Ig conjugate, 1:50 dilution
Anti-mouse Ig conjugate, 1:50 dilution
TMB substrate solutions A and B (commercially available from Sigma Chemical Co)

96 well microtiter plate (Immulon 1, Dynatech Laboratories, Inc. or equivalent)
Pipetman, P-200
Multichannel pipet (optional)
Disposal pipet tips
Plastic squirt bottle
Test tubes, glass, 12 x 75
Test tubes, plastic, 15 ml and 50 ml
Test tube racks
Vials or microcentrifuge tubes
5 ml pipets
Vortex mixer (optional)
ELISA microtiter plate reader (BIO-TEK Instruments or equivalent); optional and can be replaced by spectrophotometer with Microcuvettes

Pre-lab Preparation

1. Phosphate-buffered Saline (PBS), 10X
   2.3 g $NaHPO_4$ (anhydrous) (1.9 mM)
   11.5 g Na_2HPO_4 (anhydrous) (8.1 mM)
   90.0 g NaCl (154 mM)
   Add distilled H_2O to 800 ml
   Adjust to desired pH (7.2 to 7.4)
   Using 1 M NaOH or 1 M HCl
   Add H_2O to 1 liter

2. Carbonate buffer
   1.6 g Na_2CO_3 (15 mM)
   2.9 g $NaHCO_3$ (35 mM)
   0.2 g NaN_3 (3.1 mM)
   Add distilled H_2O to 1 liter
   Adjust to pH 9.5

   Caution: Sodium azide is poisonous; wear gloves.

3. PBS (1X) with 3% milk
   Dissolve 3 g nonfat milk powder in 100 ml PBS. Prepare during the day of use.

4. 1% Bovine serum albumin solution, (Antigen I)
   Dissolve 0.1 g BSA (e.g. from Sigma Chemical) in 10 ml carbonate buffer.
   Alternatively, provide a vial with 0.1 g BSA.

5. 1% Gelatin solution, (Antigen II)
   Dissolve 0.1 g gelatin (e.g. from Sigma Chemical) in 10 ml carbonate buffer.
   Alternatively, provide a vial with 0.1 g gelatin.

6. Rabbit anti-BSA antiserum 1:10 (Rabbit, Serum A)
   Mix 0.1 ml of rabbit anti-BSA antiserum (e.g. from Sigma Chemical) with 0.9 ml of 1X PBS. Prepare 0.1 ml aliquots in vials and freeze.

7. Normal rabbit serum 1:10 (Rabbit serum B).
   Prepare as above. Keep aliquots (0.1 ml) frozen.

8. Anti-Rabbit Ig Conjugate, 1:50
   Obtain goat anti-rabbit Ig conjugated with horseradish peroxidase (e.g. from Jackson ImmunoResearch Lab, Inc.)
   Mix 0.1 ml with 4.9 ml 1X PBS with 3% milk. Prepare 0.2 ml aliquots and freeze.
   (Adjust dilution according to batch and vendor's suggestions.)

Pre-lab Preparation, continued

9. Anti-Mouse Ig conjugate 1:50

   Obtain rabbit anti-mouse Ig conjugated with horseradish peroxidase and prepare dilution as above. Keep aliquots (0.2 ml) frozen.

10. TMB Substrate Solutions*

    Solution A

    In a one liter volumetric flask dissolve 1.0 g Tetramethylbenzidine (TMB) (Sigma Chemical) 600 mls methanol. Bring to 1.0 liter with glycerol. Store at 4°C in foil wrapped bottle.

    Solution B

    In a one liter volumetric flask dissolve in 500 ml distilled H_2O
    22.82 g K_2HPO_4
    19.2 g Citric Acid
    1.34 mls 30% H_2O_2

    Bring to 1.0 liter with distilled H_2O. Thimerosal (0.01%) may be used as a preservative.

* Also available ready-to-use from Sigma Chemical

Method

Lab Period 1

Antigen Application to Plate

1. Prepare antigen I (BSA) and antigen II (gelatin) as described in C-6.

2. Copy the chart (Figure 2) below. The numbers and letters on the chart correspond to the wells in the ELISA test plate. Along the sides of the chart record the materials added to each well as you continue with the procedure.

Method, continued

	1	2	3	4	5	6	7	8	9	10	11	12
A												
B												
C												
D												
E												
F												
G												
H												

Figure 2. Record Chart

3. Using a pipetman, add 100 µl of Antigen I to wells in rows A, B, C, and D, columns 1 through 5 and 8 through 12 of the ELISA plate. Leave columns 6 and 7 empty to separate the two sections. (If a pipetman is not available, use a dropping pipet and add 3 drops of antigen solution.)

4. Using a clean pipet tip, add 100 µl of Antigen II to wells in rows E, F, G, and H, columns 1 through 5 and 8 through 12 of the ELISA plate. Record on your chart what was added to each row.

5. Seal the ELISA plate with plastic wrap and incubate at 37°C for 1 hour. (Room temperature incubation is acceptable.) Refrigerate overnight.

Antibody Dilution in Tubes

6. Prepare stock solutions Rabbit Serum A (label tube #1) and Rabbit Serum B (label tube #2) as described in C-6.

7. Label 8 15 ml test tubes: 2, 3, 4, 5, 9, 10, 11, 12.

8. Using a 5 ml pipet, add 4 ml of phosphate buffered saline with dry milk (PBS:3% milk) to each of the 8 test tubes.

Method, continued

9. Using test tube 1 containing Rabbit Serum A, perform a serial dilution using a 5 ml pipet as follows:

   a. Transfer 4 mls of rabbit serum A from tube 1 to tube 2. Mix well by drawing liquid up and down several times.

   b. Transfer 4 ml from tube 2 to tube 3. Mix well.

   c. Transfer 4 ml from tube 3 to tube 4. Mix well.

   d. Transfer 4 ml from tube 4 to tube 5. Mix well.

10. Using tube #8 containing rabbit serum B, repeat the serial dilution as described above for tubes 9 through 12. Seal all of the tubes and refrigerate until needed (step 14).

Lab Period 2

Antibody Addition to Plate

11. Empty contents of ELISA plates into a sink by turning upside down and flicking. Blot on paper towels.

12. Fill each well on the ELISA plate with PBS: milk to the top of the wells. Let set for 15 minutes. Empty as above. (This step serves the purpose of saturating the nonspecific binding sites in each well.)

13. Wash by flicking the plate and refilling wells with PBS only. Let sit for 2 minutes. Empty and blot as above. Repeat.

14. Using a pipetman, add 100 µl of diluted rabbit serum to each well according to the chart below. The tube number corresponds to the row on the ELISA plate. Use a clean pipet tip for each rabbit serum (A and B). Record all information on the chart in fig. 1. (If a pipetman is not available, use a dropping pipet and add 2 drops of each serum preparation.)

15. Cover plate and incubate for one hour at room temperature.

16. Empty ELISA plates into sink and blot on paper towels.

Method, continued

ELISA Plate Set Up

Tube #	Rabbit Serum	Dilution	Column	Row
1	A	1:1000	1	A - H
2	A	1:2000	2	A - H
3	A	1:4000	3	A - H
4	A	1:8000	4	A - H
5	A	1:16000	5	A - H
8	B	1:1000	8	A - H
9	B	1:2000	9	A - H
10	B	1:4000	10	A - H
11	B	1:8000	11	A - H
12	B	1:16000	12	A - H

17. Wash as in step 13. Wash a total of three times.

18. Add 100 µl of Rabbit Conjugate to wells A and B, E and F, rows 1 through 5 and 8 through 12. Add 100 µl of Mouse Conjugate to wells C and D, G and H, rows 1 through 5 and 8 through 12. Cover plates. Record information on chart.

19. Incubate for 15 minutes at 37°C

Substrate Addition to Plate

20. Prepare TMB substrate as described in C-6. Working solution of substrate should be freshly prepared right before use.

21. Empty plates and wash as described in steps 16 and 17 above.

22. Add 200 µl of fresh substrate to wells in rows A, B, C, D, E, F, G, and H, columns 1 through 5 and 8 through 12. A color change (colorless to blue) after few minutes indicates a positive antigen-antibody reaction.

Method, continued

23. Read optical density or absorbance at wavelength 650 nm on a ELISA plate reader after 15 and 30 minutes. Alternatively, transfer the reaction product from each well to a microcuvette and read OD_{650} in a spectrophotometer. If neither a plate reader nor a spectrophotometer is available, place ELISA plate on an index card or sheet of white paper and observe any color change. Make observations between 15 and 30 minutes and score according to the following:

    +++ very reactive
    ++ moderately reactive
    + slightly reactive
    - no reaction

24. Record your results on your chart.

Results

1. Prepare a table as shown below and summarize your results. Give a title to your table.

	BSA		GELATIN	
	Rabbit Conjugate	Mouse Conjugate	Rabbit Conjugate	Mouse Conjugate
Anti-BSA				
1:1000				
1:2000				
1:4000				
1:8000				
1:16000				
Normal Serum				
1:1000				
1:2000				
1:4000				
1:8000				
1:16000				

2. Using the data in your table above, prepare a graph to compare the activities of anti-BSA and normal serum control. Give a title to your figure. (Which format, table and figure, do you think is more effective in showing your results?)

Results, continued

3. Write a paragraph to interpret your data. It is important here to consider the results you obtain with both your test system (i.e., anti-BSA, BSA and Rabbit Conjugate) and controls (normal serum, gelatin and mouse conjugate). What can you say about the specificity and potency (i.e., titer) of the anti-BSA preparation?

4. Discuss any problem you may have encountered in your experiment as well as any discrepancies you may have found in your results.

5. Try to draw a conclusion of your experiment. Do you think you achieved the goal of your experiment? Explain.

Results Analysis

1. A typical and expected result is shown in the chart below. Actual result may vary according to the reagents used.

<table>
<tr><td></td><td colspan="4">|<—Rabbit Serum A——>|</td><td></td><td colspan="4">|<—Rabbit Serum B———>|</td><td></td><td></td><td></td></tr>
<tr><td></td><td colspan="4">Antigen I (1% BSA)</td><td></td><td></td><td></td><td></td><td></td><td></td><td></td><td></td></tr>
<tr><td></td><td>1</td><td>2</td><td>3</td><td>4</td><td>5</td><td>6</td><td>7</td><td>8</td><td>9</td><td>10</td><td>11</td><td>12</td></tr>
<tr><td>R-A</td><td>+++</td><td>++</td><td>++</td><td>+</td><td>-</td><td></td><td></td><td>-</td><td>-</td><td>-</td><td>-</td><td>-</td></tr>
<tr><td>R-B</td><td>+++</td><td>++</td><td>++</td><td>+</td><td>-</td><td></td><td></td><td>-</td><td>-</td><td>-</td><td>-</td><td>-</td></tr>
<tr><td>M-C</td><td>-</td><td>-</td><td>-</td><td>-</td><td>-</td><td></td><td></td><td>-</td><td>-</td><td>-</td><td>-</td><td>-</td></tr>
<tr><td>M-D</td><td>-</td><td>-</td><td>-</td><td>-</td><td>-</td><td></td><td></td><td>-</td><td>-</td><td>-</td><td>-</td><td>-</td></tr>
<tr><td colspan="13">Antigen II (1% gelatin)</td></tr>
<tr><td>R-E</td><td>-</td><td>-</td><td>-</td><td>-</td><td>-</td><td></td><td></td><td>-</td><td>-</td><td>-</td><td>-</td><td>-</td></tr>
<tr><td>R-F</td><td>-</td><td>-</td><td>-</td><td>-</td><td>-</td><td></td><td></td><td>-</td><td>-</td><td>-</td><td>-</td><td>-</td></tr>
<tr><td>M-G</td><td>-</td><td>-</td><td>-</td><td>-</td><td>-</td><td></td><td></td><td>-</td><td>-</td><td>-</td><td>-</td><td>-</td></tr>
<tr><td>M-H</td><td>-</td><td>-</td><td>-</td><td>-</td><td>-</td><td></td><td></td><td>-</td><td>-</td><td>-</td><td>-</td><td>-</td></tr>
</table>

+++ = very reactive
++ = moderately reactive
+ = slightly reactive
- = not reactive

M = mouse conjugate

R = rabbit conjugate

1:4000 = the highest dilution of anti-BSA antibody that gives a visible antigen-antibody reaction.

Results, continued

2. The end result of an ELISA test may be subjectively assessed by eye to provide a "yes" or "no" answer. Positive results are indicated by a color reaction which can be distinguished from the colorless (or very faint color) wells of negative controls.

   Preferably, the colored reaction product should be read in a ELISA microplate reader or a spectrophotometer to measure more precisely its absorbance (i.e. optical density) at the maximal absorption wavelength. A threshold value is determined (usually the upper limit of absorbance value shown by the negative controls) and test samples giving an absorbance value above the threshold level (e.g. 2 to 3 x negative control) is considered as positive.

   Since the speed of enzyme reaction may vary the end result should be read as soon as possible, and every 5-10 minutes thereafter for 30 minutes. If the reaction appears too fast it can be terminated by adding 50 μl of hydrofluoric acid (HF, 1:400 dilution) to each well.

   The result may be expressed as either endpoint titration or a quantitative O.D. measurement, or as both.

3. The procedure will work for most commercial reagents (e.g. the antigens, rabbit sera and enzyme conjugates) but the concentrations may need to be adjusted by following vendor's recommendations.

4. Incubation times with antibody can be reduced to 30 minutes if speed is important although some loss of binding may occur. Incubation at 37°C will speed the reaction but may also increase nonspecific binding leading to high background

5. No positive results - sources of problem

   a. BSA antigen: check concentration and date of antigen solution; check plates that are used for coating; check antigen coating buffer.

   b. Rabbit Anti-BSA serum - check date and dilutions of antibody solutions; increase concentration of antibody; call supplier for advice.

   c. Anti-rabbit Ig conjugate: check date and dilution of conjugate; increase concentration of conjugate; call supplier for advice.

   d. TMB substrate: Prepare fresh working solution; check stock solutions A and B.

Results, continued

e. Check (c) and (d) by mixing 100 µl of diluted conjugate with TMB working solution. Color reaction should result.

6. High Background - sources of problem

   a. Plate not-saturated: check saturation buffer (PBS milk); prolong saturation time; increase milk concentration; try other carrier proteins such as 10% normal goat serum.

   b. Reaction too fast or end results read too late: try reading the results soon after substrate addition; stop reaction with HF.

   c. Conjugates too concentrated: Dilute conjugates further; call supplier for advice.

   d. Sera too concentrated. Dilute antiserum and normal serum further; call supplier for advice.

UNIT 5: ANTIBODY SPECIFICITY AND DIVERSITY

MODULE 19: RADIAL IMMUNODIFFUSION

Dennis Bogyo

* Introduction
* Safety Guidelines
* Experimental Outline
* Materials
* Pre-lab Preparation
* Method
* Results

Introduction

Radial immunodiffusion is a technique that can quantitatively determine the level of an unknown antigen. Preformed antibody is incorporated into liquid agar which is poured into a Petri dish and allowed to gel. Small wells cut into the agar dish are filled with solutions of known concentration antigen and an unknown solution sample. The antigen solution diffuses outwards from the well in a circular pattern. In the endpoint or Mancini method used in this module, antibody is present in excess amount and diffusion continues until a stable ring of antigen-antibody precipitate forms, generally within 24 to 48 hours.

For each standard antigen concentration an endpoint precipitation ring of a certain diameter is formed. The diameter squared measurements of the rings plotted against the known concentration of each starting standard yields a straight line. From this linear calibration curve the concentration of the unknown antigen sample may be determined. Unlike many gel and liquid precipitation techniques which qualitatively detect antigen, RID is a sensitive quantitative technique that is often used clinically to detect patient levels of blood proteins.

Safety Guidelines

No mouth pipetting.

Students will wear safety glasses.

Dispose of used RID plates through proper laboratory waste disposal procedures.

Experimental Outline

Most kits are designed to be performed quickly, and generally take 10-30 minutes to complete. Incubations to allow for diffusion and ring formation generally occurs within 24 to 48 hours. This lab can be included in another related lab either to illustrate applications of antibodies to start the lab or during an incubation or other waiting period.

Perform test: follow test instructions - generally 10-30 min.

Analyze results: interpret test results by measuring ring diameters- about 15 min.

Materials

RID plates
Set of reference standard solutions for antigen
Calibrated ruler, mm.
Pipette gun, 5 microliter or plastic micropipette
Graph paper

A kit, which contains all the critical reagents required for this experiment, is available from EDVOTEK (1-800-EDVOTEK).

Pre-lab Preparation

Generally none, other than obtaining kits and samples.

Method

1. Perform the test using the samples provided, following the directions provided with the kit.

   a. Add 5 microliters of antigen solution standards to each of four wells. Rinse micropipettes or change tips.

Method, continued

b. To a fifth well add 5 microliters of the unknown antigen solution. Snap the cover of the Petri dish back on. Place the dish into a zip lock bag and seal.

c. Allow the dish to sit undisturbed for 24-48 hours at room temperature on a flat surface.

d. After precipitin rings have reached equilibrium, remove the plate from the bag and place on a light box or back lit plate.

2. Record the results of the test.

   a. Measure the diameter of each precipitin ring with the ruler to the nearest tenth of a millimeter.

   b. Square the ring diameters and plot these values on the Y axis of the graph paper versus the known starting antigen concentrations.

   c. Draw the best fit line through the standard points.

   d. Using the standard line determine the value of the unknown antigen.

Results

1. Interpret the results of the test, considering any control sample run.

2. Plot the values of precipitin ring diameter squared versus known antigen concentrations. Draw the best fit line through these points. Calculate the value of the unknown antigen concentration from this line.

UNIT 5: ANTIBODY SPECIFICITY AND DIVERSITY

MODULE 20: ANTIBODIES AS TOOLS: USING COMMERCIALLY AVAILABLE KITS

Karen K. Klyczek

* Introduction
* Safety Guidelines
* Experimental Outline
* Materials
* Pre-lab Preparation
* Method
* Results

Introduction

Immunoassays are tests utilizing interactions between antibodies and antigens, and can be used to detect and quantitate either an antigen or an antibody. These assays are powerful tools used widely in medicine, agriculture, and food services, and in many fields of basic research. Three general classes of immunoassays and some common variations are described briefly below.

Antibody capture assays can be used to detect and quantitate either antigens or antibodies. The general protocol involves immobilizing an antigen on a solid phase (e.g. a well of a plastic dish or a membrane), and then adding antibody and allowing it to bind to the immobilized antigen. The solid phase is then washed to remove any unbound material. To detect antibody binding, the antibody can be labeled directly, or a labeled secondary reagent, which will specifically recognize the antibody, can be used (methods of labeling are discussed below). The amount of antibody that is bound determines the strength of the signal.

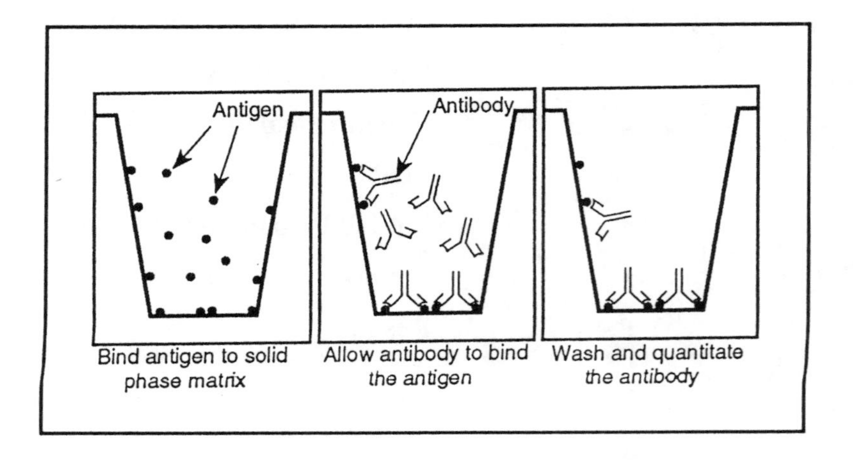

Fig. 1 Antibody capture assay

Antigen capture assays are used primarily to detect and quantitate antigens in a sample. Unlabeled antibody is immobilized on the solid phase, and labeled antigen is allowed to bind to the immobilized antibody. The most common variations on this method involve competition between a known quantity of labeled antigen and unlabeled antigen in a test sample. When these two samples are mixed and added to the immobilized antibody, they will compete for binding to the antibody binding sites. If the test sample contains a high concentration of the antigen, it will compete effectively with the labeled antigen and little labeled antigen will bind. Thus, the amount of label present (i.e. the signal detected) will be inversely proportional to the amount of antigen present in the test sample.

Introduction, continued

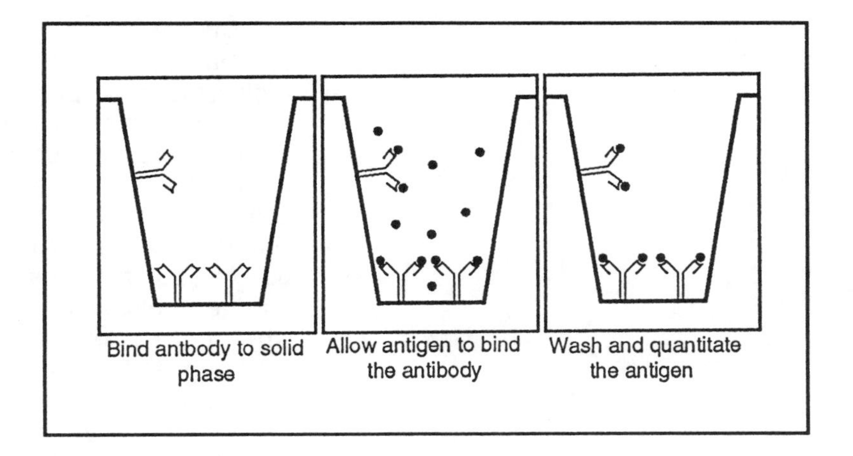

Fig. 2 Antigen capture assay

Two antibody (sandwich) assays are used primarily to determine the antigen concentration in unknown samples; the assays are quick and accurate. The assay requires two antibodies that bind to non-overlapping regions of the same antigen. One antibody is immobilized to the solid phase, and the antigen in the test sample is allowed to bind. Unbound material is washed away, and a labeled, second antibody is added which will bind to any antigen which bound to the immobilized antibody. The major advantages of this technique are that it is very sensitive and specific, and crude antigen samples can be used.

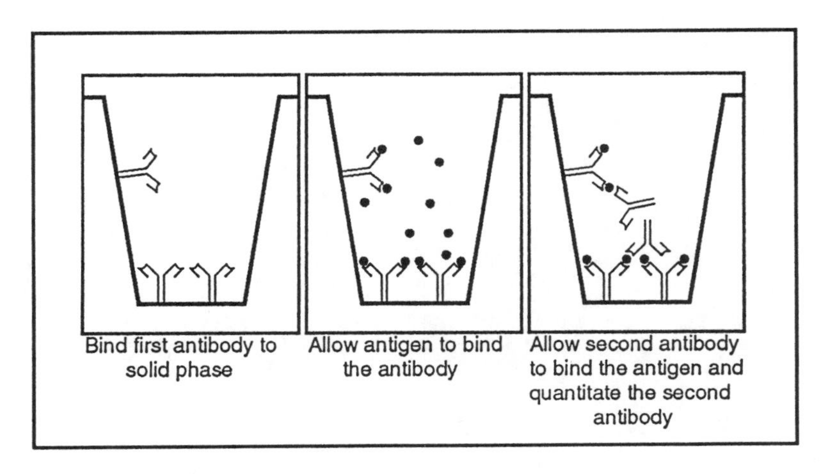

Fig. 3 Two-antibody sandwich assay

Introduction, continued

All immunoassays rely on labeled antigens, antibodies, or secondary reagents for detection. These proteins can be labeled with radioactive molecules (usually radioactive iodine), enzymes with color-producing substrates, or fluorescent dyes. Of these, radioactive labeling can be used for almost all types of immunoassay; it is easy to detect and quantitate. However, it is not always desirable to use radioactive materials due to safety and disposal considerations. Enzyme-labeled reagents yield semi-quantitative results and are now very commonly used, particularly in clinical labs and other uses outside of research labs. A colorless substrate for the enzyme, which turns color when acted upon by the enzyme, is used for detection. An immunoassay using an enzyme label is usually called an ELISA (Enzyme-Linked Immunosorbent Assay).

This module will examine immunoassays which are used commercially and which are readily available as kits. All utilize enzyme labels. The immunoassays will be performed according to the instructions provided with the kit. Using the information provided in this introduction and in the kit instructions, the type of immunoassay represented by the kit, and the nature of the interaction between antibodies and antigens in the assay will be determined.

Safety Guidelines

Standard lab safety procedures should be followed. Note any special safety precautions described in any kit instructions used.

No special precautions are required, unless indicated for a particular kit or if human biological samples are used. The latter should be disposed of as biohazard waste.

Experimental Outline

Timetable of Events

Most kits are designed to be performed quickly, and generally take 10-30 minutes to complete. This lab can be included in another related lab either to illustrate applications of antibodies to start the lab or during an incubation or other waiting period. Alternatively, several kits can be performed and analyzed to fill a 2- or 3-hour lab period.

Perform test; follow test instructions - generally 10-30 min.

Analyze results. Interpret test results as positive or negative and determine location of antibodies and antigens, mechanism of labeling - about 15 min.

Materials

Immunoassay kit and instructions
Test sample and/or control samples, as indicated by instructor

Antibody based test kits, such as home pregnancy kits, are available in drug stores. These kits illustrate how simplified the use of antibodies as tools has become. Many other types of kits are available commercially.

Test samples and controls. Control urine samples (containing high and low levels of human chorionic gonadotrophin) are available commercially for pregnancy test kits. These purchased controls are usually preferable to obtaining urine samples from volunteers.

Most drug stores, grocery stores, and discount stores carry a large selection of home pregnancy test kits. One type that is easy to explain is the brand First Response, which is set up as described

Materials, continued

in the answer to question 1 above, with all of the components added in separate steps. Many other brands are set up with all reagents in a single test unit, and the urine is drawn by capillary action past all of the reagents. Other sources of test kits:

Fisher Scientific
711 Forbes Avenue, Pittsburgh, PA 15219; 214-562-8300
Carries control hCG urine samples for pregnancy tests, and a variety of clinical test kits.

BioMetallics, Inc.
Princeton, NJ; 1-800-999-1961
TARGET milk progesterone test kit
If possible, obtain fresh milk samples from a local farm. Milk from the grocery store yields a very low or negative results; this milk is usually obtained from a farm bulk tank so it tends to have an average level of progesterone, but apparently some antigenic activity is lost when the milk is homogenized. If you are in a rural area, a local farm supply store may carry similar test kits.

Agri-Diagnostics Assoc.
One Executive Drive, Moorestown, NJ 08057; 800-322-KITS
REVEAL turf grass disease detection kits for identifying specific fungal pathogens; kits come with positive control samples, and test samples can be obtained from local golf courses. Be sure kits and control samples have not reached expiration dates.

Pre-lab Preparation

Generally none, other than obtaining kits and samples

Method

Perform the test using the samples provided, following the directions provided with the kit. Record the results of the test.

Results

Interpret the results of the test, considering any control sample run. What valid conclusion can be drawn from the test?

UNIT 6

Cellular Immunology

UNIT 6: CELLULAR IMMUNOLOGY

MODULE 21: IDENTIFICATION OF LYMPHOID CELLS IN BLOOD SMEARS AND TISSUE SECTIONS

Dennis Bogyo

- Introduction
- Safety Guidelines
- Experimental Outline
- Materials
- Pre-lab Preparation
- Method
- Results

Introduction

Blood cells undergo progressive differentiation from stem cells originating in the bone marrow. In the presence of differentiation inducing growth factors B lymphocytes mature in the bone marrow and T lymphocytes mature in the thymus; both of these lymphoid organs are classified as primary lymphoid organs. Peripheral blood, lymph nodes, the spleen, and the tonsils are classified as secondary lymphoid organs since they contain mature, fully functional lymphocytes. By comparing cells from the bone marrow with cells from the peripheral blood and the lymph nodes some of the cytological differences between immature stem cells and mature lymphocytes may be observed.

The bone marrow contains lymphoid and myeloid cells at various stages of differentiation. The larger more primitive cell types show less dense nuclei and are more closely related to immature stem cells. These undifferentiated cells are nonfunctional and differences between lymphoid and various myeloid cell types are more difficult to distinguish.

Peripheral blood smears show well defined different cell types. Functional myeloid cells include erythrocytes, granulocytes, monocytes and platelets. Lymphocytes show a circular dark staining nucleus under normal conditions. It is not possible to distinguish T lymphocytes from B lymphocytes in peripheral blood using standard Wright's-Giemsa stain.

Distinct circular regions of cells in the lymph nodes called germinal centers are present near the outer, cortical edge of the organ. These centers represent sites where antigen processing cells like macrophages have stimulated B lymphocytes with antigen and induced clonal expansion of those B lymphocytes that were able to respond to antigen. T lymphocytes, including key T helper cells, are located in the lymph node in the paracortical region that lies closer to the central portion of the lymph node.

There may be a need to use standard definitions for the histology and cellular terms. A proposed set of definitions follows:

B-cell: Lymphocyte that is the key cell in the production of antibodies following stimulation

Bone Marrow: Source of stem cells for all cellular components of blood.

Clonal Expansion: The stimulation of a single cell to undergo cell division and produce clonal descendants

Cortical: Outer layer or region of an organ or tissue

Introduction, continued

Differentiation: The process of selective gene activation as a cell progresses in development from a stem cell

Erythrocyte: Differentiated red blood cells

Germinal Centers: Areas of lymphoid tissue where stimulated lymphocytes are dividing

Granulocytes: Phagocytic white cells that show prominent cytoplasmic granules and are abundant in peripheral blood

Lymph Nodes: Secondary lymphoid organs along length of lymphoid vessels that act as sites for antigen trapping and B cell stimulation

Macrophage: A white cell derived from a monocyte that helps process and present antigen

Monocyte: A white cell type found in peripheral blood that, like more abundant granulocytes, has phagocytic activity

Myeloid Cells: A family of cells which develop from a bone marrow stem cell to become erythrocytes, monocytes, granulocytes, and platelets

Paracortical Region: A region outside of the cortex of an organ or tissue, often farther from the external edge

Spleen: A secondary lymphoid organ which stores white cells and red blood cells

Thymus: A primary lymphoid organ which is essential for the differentiation of T lymphocytes

Tonsils: Secondary lymphoid organ that stores differentiated lymphocytes

T Lymphocyte: Lymphocyte that is a key cell in stimulating immune cells and forming T killer cells for rejecting grafts and tumors

T helper cell: Type of T lymphocyte that helps stimulate both functional T and B cells through production of soluble effectors called lymphokines

Safety Guidelines

Use commercial slides and do not make peripheral blood smears using finger sticks.

Clean the oil immersion lens after use by using only microscope grade lens tissue.

Before trying to remove slides from the microscope stage rotate back to low power lens to provide clearance.

Experimental Outline

1. Identify the different mature cell types in peripheral blood.
2. Identify different immature cell types in the bone marrow and compare these to peripheral blood.
3. Identify the overall structure of the lymph node and the location and morphology of lymphoid cells in the organ. Distinguish T from B lymphocytes based on their location.

Materials

Commercial slides of peripheral blood, Wright's-Giemsa stained.

Commercial slides of lymph nodes, Hematoxylin-Eosin stained.

Commercial slides of bone marrow smears, Wright's stain, or white bone marrow, Hematoxylin-Eosin stained.

Microscopes with oil immersion lenses

Immersion oil

Microscope lens tissue

Pre-lab Preparation

1. Obtain slide sets of peripheral blood, bone marrow, and lymph nodes, microscopes with oil immersion lenses, immersion oil, and microscope lens cleaning tissue.
2. During pre-lab discussion draw characteristic lymphoid cells and general architecture of the tissues.
3. Instruct students in the proper sequence of steps in using the oil immersion lens.

Method

1. For peripheral blood smears move the slide under high power to the feather edge end of the smear where the lower density of cells does not lead to clumping. Switch to oil immersion and make minor fine focus changes after contact of the lens with the oil drop on the slide. Draw the cell types visible.
2. Follow a similar procedure for bone marrow smears and draw the cell types observed.
3. For tissue sections get an overall orientation of the tissue regions at low power magnification. Proceed up through high power and then to oil immersion lenses. Draw the cell types present. For lymph nodes draw cells seen in the germinal centers and in the paracortical region.

Results

Compare the drawings you have made from the three different lymphoid organs.

UNIT 6: CELLULAR IMMUNOLOGY

MODULE 22: SPLEEN CELL PREPARATION FROM MOUSE

Dennis Bogyo

* Introduction
* Safety Guidelines
* Experimental Outline
* Materials
* Pre-lab Preparation
* Method
* Results

Introduction

The preparation of lymphocytes from mouse spleen is a technique which readily provides a source of functional T cells, B cells, and macrophages. Unlike the thymus and bone marrow which contain undifferentiated lymphoid and myeloid cells, the spleen is classified as a secondary lymphoid organ along with the lymph nodes and the peripheral blood and contains active differentiated cells. The spleen is easily dissociated into a single cell suspension and provides a high yield (approximately 1-2 x 10^8 cells per spleen).

Cells obtained from the spleen are an excellent starting material for primary cell cultures. Cells obtained from the spleen may be fractionated into T lymphocytes, B lymphocytes, and macrophages and their activities studied using in vitro systems. Historically this approach along with controlled reconstitutions after fractionation led to discovery of the nature of T cell, B cell, and macrophage cooperation.

After preparing and counting splenic lymphocytes an investigator may proceed with structural or functional studies of the cells. Splenic lymphocytes from immunized animals are key starting cells for the production of hybridomas and monoclonal antibodies. The activity of T lymphocytes sensitized in culture to react against tumor cells is being studied to determine more effective treatments against specific types of cancer.

Instructors should obtain clearance from the animal use committee of their institution for demonstrating that they are complying with accepted federal laboratory procedure guidelines.

If mice are unavailable bovine spleen may be obtained from a slaughterhouse. The fresh tissue should be placed in BSS on ice. Some lymphoid cell lines are available from American Type Culture Collection (ATCC).

Safety Guidelines

1. No mouth pipetting.
2. No ethanol beakers near open flames .
3. Students will wear safety glasses.

Experimental Outline

Sacrifice mice

Remove spleen

Sieve spleen cells through wire mesh

Wash cells

Viable cell count with trypan blue

Cell count with hemocytometer

Materials

Balanced Salt Solution (BSS)
Acetic Acid Counting Solution
Trypan Blue Vital Dye Solution
Beaker Containing 70% Ethanol
Squeeze Bottle Containing 70% Ethanol
Small Surgical Scissors
Large Surgical Scissors
Curved Hemostat
Curved forceps
Rubber Policeman
Stainless Steel Wire Mesh
Glass Centrifuge Tube, 30 ml.
Glass Tubes, 5 ml.
Glass Petri Dish
Stainless Steel Wire Mesh
Hemocytometer
Pasteur Pipettes With Rubber Bulb

Pre-lab Preparation

1. Make up 10X stocks of Balanced Salt Solution (BSS)

Stock #1	Dextrose	10 gm
	KH_2PO_4	0.6 gm
	0.5% phenol red solution	20 ml

Dissolve and bring up to 1000 ml with distilled water

Stock #2	$CaCl_2 \cdot H_2O$	1.86 gm
	KCl	4.0 gm
	NaCl	80.0 gm
	$MgCl_2$	1.04 gm
	$MgSO_4 \cdot 7H_2O$	2.0 gm

Dissolve and bring up to 1000 ml with distilled water.

Mix 10 ml of stock #1 and 10 ml of stock #2, bring up to 100 ml with distilled water. The pH should be 7.2-7.4.

2. Sterilize all instruments and wire mesh by storing in 70% ethanol. Place in a loosely covered sterile Petri dish and allow to air dry before use. Pasteur pipettes and glass tubes may be purchased sterilized. Alternatively, all materials can be wrapped in aluminum foil and autoclaved (check any plastic ware for stability under heating). Standard microbiology flaming technique can also be used for instruments.

3. Make up the following solutions for cell counting:

   0.5% acetic acid in distilled water (v/v) for lysing red blood cells
   trypan blue 10x stock, 1% (w/v) in distilled water

4. Sacrifice mice just before the laboratory in a closed chamber containing 30% CO_2. An alternative, but less-preferred method, is to use cervical dislocation.

Method

1. After the mouse skin has been wet with 70% ethanol, place the left side facing up on the paper towels.

2. Make a cut through the loose skin by pulling gently upwards and using the blunt scissors on the skin flap. This will expose the peritoneal wall.

Method, continued

3. Pull gently in opposite directions on the two sides of the skin incision to expose a wider area of the peritoneal wall.

4. With the small surgical scissors make a cut over the spleen to expose it through the peritoneum.

5. Grasp the spleen with the forceps. Pull up gently and cut away attached connective tissue which appears white using the small surgical scissors.

6. Grasp a folded corner of the stainless steel wire mesh with the hemostat. Insert the mesh into the Petri dish and add about 10 ml of BSS.

7. Put the removed spleen on the wire mesh. With a rubber policeman, rub the spleen back and forth over the mesh. Continue until all the spleen dissociates leaving only white connective tissue on the mesh surface.

8. With a sterile Pasteur pipette bring the cell suspension up and down several times to dissociate large cell clumps. Pipette the cell suspension into the glass centrifuge tube.

9. Allow the cells to settle for 5 minutes to remove larger debris. Use the Pasteur pipette to transfer roughly 2 ml volumes of cell suspensions to 5 ml glass tubes.

10. Centrifuge at 200 x g for 10 minutes. Pour off the supernatant.

11. Resuspend the cell pellet in 2 ml of BSS by vortex mixing or up and down action in a Pasteur pipette.

12. Add .1 ml of trypan blue solution to 1.8 ml of acetic acid solution. Add .1 ml of the cell suspension and mix thoroughly.

13. Fill a hemocytometer chamber with a drop of your mixture and proceed as if doing a white blood cell count. Count both total cells and the fraction which are dead (those that have taken up the blue dye).

14. Remember that a dilution factor of 20 and a volume factor of 10,000 need to be multiplied by the average count of a 4 by 4 square grid. For example, 50 cells in a 4 by 4 grid would yield a cell count of 10 million cells/ml.

Results

Record your grid counts. Average four separate grids. Record the number and per cent of dead cells per grid and average these trypan blue positive counts.

UNIT 6: CELLULAR IMMUNOLOGY

MODULE 23: IDENTIFICATION OF LYMPHOCYTE POPULATIONS

Karen K. Klyczek

Introduction to Module 3

Lymphocytes are one of the major classes of white blood cells involved in the immune response. These cells are derived from stem cells in the bone marrow, which migrate to the lymphoid organs and develop into either B lymphocytes or one of the subclasses of T lymphocytes. Each type of lymphocyte plays a distinct and critical role in immunity. These functions are mediated in part by unique proteins with specific binding properties, called receptors, present on the surface of each type of lymphocyte. For example, B lymphocytes have cell surface immunoglobulin, or antibody, molecules which bind with a specific antigen. This binding stimulates activation and multiplication of the B lymphocyte. T lymphocytes also have cell surface, antigen-specific receptors which are distinct from immunoglobulin. Another example is the Fc receptor, found on certain lymphocytes, which binds the constant region of antibody molecules, called the Fc region.

If one examines a mixed population of lymphocytes microscopically, the cells are generally indistinguishable with respect to size, shape, etc. However, reagents have been developed which will identify the specific cell surface receptors so that they are visible by microscopy, and therefore allow visual identification of specific cell types. The two experiments described below demonstrate different reagents used to identify specific subpopulations of lymphocytes within a heterogenous population of lymphocytes, such as blood cells or spleen cells.

Introduction- Experiment A

Experiment A - Identification of Fc Receptor-bearing Cells

Most mature B lymphocytes, as well as other accessory cells such as macrophages, possess cell surface receptors for the Fc portion of immunoglobulin, or antibody, molecules of the IgG class. Other cell types have receptors for the Fc region of other immunoglobulin classes. The Fc region of an antibody is also called the constant region, since all antibodies of a particular class have the same Fc region. In contrast, the Fab portion of an antibody contains the highly variable regions of the protein, where the specific antigen combining sites are formed (Figure 1).

The role of Fc receptors in the immune function of B lymphocytes remains unclear. On other immune cell types, however, Fc receptors may facilitate the destruction of cells or particles coated by antibodies during an immune response. It has been demonstrated that Fc receptors on macrophages promote binding and subsequent engulfing and destruction of antibody-antigen complexes

Introduction - Experiment A, continued

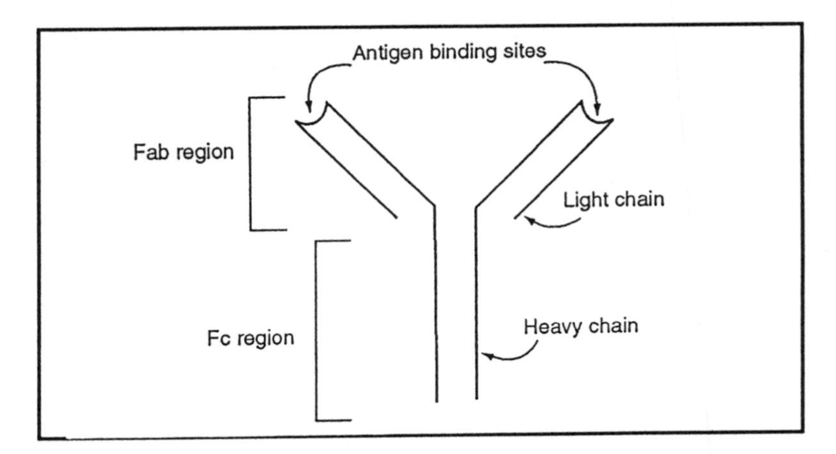

Figure 1. General structure of an antibody molecule

which would damage host tissues if allowed to accumulate. Another important example is the Fc receptors for IgE antibodies found on basophils and mast cells, which are probably involved in allergic reactions.

Cells bearing Fc receptors on their surface can be identified easily by the formation of rosettes with antibody-coated erythrocytes (red blood cells). This procedure involves mixing a suspension of lymphocytes, for example spleen cells, with erythrocytes which have been coated with IgG antibodies produced against the erythrocytes. The Fab portion of the antibodies will attach to the erythrocyte surface. The Fc portion of the antibodies will be exposed, and will bind with the IgG Fc receptors on the B lymphocytes and other cells (Figure 2). Cells bearing Fc receptors are then

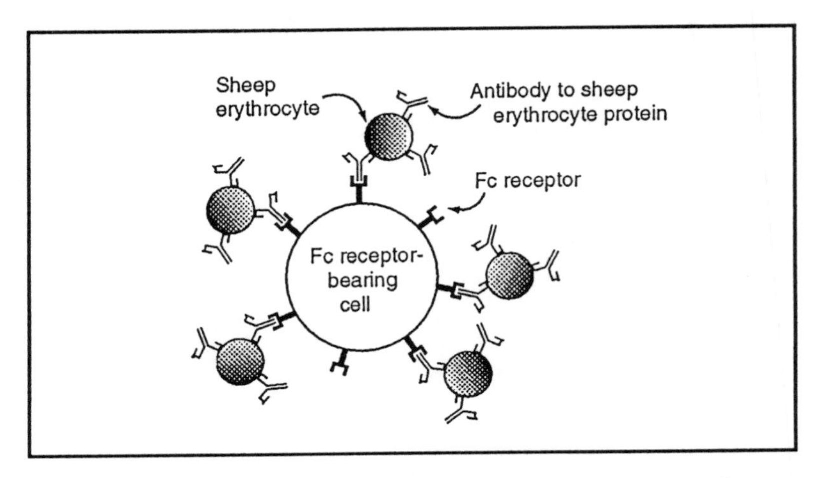

Figure 2. Formation of rosettes with antibody-coated erythrocytes and Fc receptor-bearing cells

Introduction - Experiment A, continued

identified microscopically by the cluster of erythrocytes around them, called a rosette. Rosetted cells can be separated from cells which have not formed rosettes by differential centrifugation techniques. Removal of the erythrocytes then yields a cell population enriched for Fc receptor-positive cells. Variations on this procedure are commonly used to isolate human lymphocyte populations inn clinical labs, and to isolate or deplete various white blood cell populations in research labs.

In this module, sheep erythrocytes, which have been coated with IgG antibodies produced against sheep erythrocyte stroma proteins, will be used to identify IgG Fc receptor-bearing cells in a population of cells isolated from mouse spleen. The mouse Fc receptors for IgG will bind with the Fc portions of either mouse or rabbit IgG molecules. The percentage of mouse spleen cells which have Fc receptors for IgG on their surface can then be calculated.

Safety Guidelines - Experiment A

Standard lab safety practices should be followed.

Experimental Outline - Experiment A

Timetable of Events - Experiment A

The entire lab procedure, from the mixing of lymphocytes and coated erythrocytes to cell counting, can be completed in less that one hour. This procedure may be performed in the same two to three-hour lab period as another short experiment. Alternatively, students, rather than the instructor, may perform the antibody coating of erythrocytes, or any other preparation steps, during the lab period. Spleen cells should be harvested and erythrocytes removed on the day of the lab. The sheep erythrocytes should be prepared within one week prior to the lab and stored in the refrigerator.

Pre-lab Preparation

Spleen cell preparation:	30 minutes
Removal of erythrocytes from spleen cells (hypotonic or acetic acid lysis):	30 minutes
Preparation of erythrocytes:	40 minute (55 min. if making HBSS from scratch).
Antibody coating of erythrocytes:	55 minutes

Lab Procedure

Generate rosettes: Mix spleen cells and antibody-coated erythrocytes and incubate; stain with crystal violet - about 25 minutes.

Observe and quantitate rosettes - about 15 minutes.

Materials - Experiment A

NOTE: If the use of mice for this lab is not appropriate, it is suggested that Experiment B be performed instead of this one, since either bovine spleen (from a slaughterhouse) or cultured lymphocytic cell lines can be used. It may be possible to use such cells in this experiment, but that has not been tested yet by the author.

Antiserum produced in rabbits or mice against sheep erythrocytes (1 ml)
Sheep blood or glutaraldehyde-fixed sheep erythrocytes (4 ml)

Hank's Balanced Salt Solution (HBSS) (250 ml)

Lymphocytes isolated from mouse spleen, purified by acetic acid lysis to remove erythrocytes (Unit 6, Module 2), 1 x 10^6 cells/ml (See below for alternative protocol for removing erythrocytes: hypotonic lysis)

5% suspension, in Hanks balanced salt solution (HBSS), of sheep erythrocytes coated with IgG antibodies

Crystal violet, 0.5% w/v in distilled water

17 x 100 mm clear tubes, either with caps or with parafilm for sealing

Microscope slides and cover slips

5 ml pipets

1 ml pipets (or automatic micropipet)

Pipet pumps

37°C water bath

12 x 75 mm glass tubes

Microscope capable at least 200x total magnification

Pasteur pipets and bulbs

Materials - Experiment A, continued

Sources

Sigma Chemical Co.
P.O. Box 14508
St. Louis, MO 63178; 1-800-325-3010

For indirect, fluorescent labeling:
#M7019 Goat antibody to mouse IgM
#F9259 FITC-labeled rabbit antibody to goat immunization

For indirect, enzyme labeling:
#M7019 Goat antibody to mouse IgM
#A7650 Alkaline phosphatase-labeled rabbit antibody to goat immunoglobulin

For direct, fluorescent labeling:
#F9259 FITC-labeled goat antibody to mouse IgM
Also, antibodies against IgM from other species are available if cells other than mouse are used.

American Type Culture Collection (ATCC)
12301 Parklawn Drive
Rockville, MD 20852; 1-800-638-6597

Possible cultured mouse cell lines- in place of spleen cells:

Surface IgM positive:
#ATCC 1702, WEHI-231
#ATCC CRL 1704, WEHI-279
#ATCC TIB 209, x16C8.5
(All are mouse B cell lymphomas.)

Surface IgM negative:
#ATCC TIB 47, BW5147.3
#ATCC TIB 155, LBRM-33
(Both T cell lymphomas.)

Pre-lab Preparation - Experiment A

1. Preparation of Hank's Balanced Salt Solution: This can be purchased either as a 1x or 10x solution or as a powder. If prepared in the laboratory, the following recipe will make one liter of 1x HBSS:

   Solution A:
   - 1.0 g D-glucose
   - .01 g Phenol red
   - .06 g KH_2PO_4
   - .09 g $Na_2HPO_4 \cdot 7H_2O$

   Solution B:
   - .14 g $CaCl_2$
   - 0.4 g KCl
   - 8.0 g NaCl
   - 0.1 g $MgCl_2 \cdot 6H_2O$
   - 0.1 g $MgSO_4 \cdot 7H_2O$

   0.35 g Sodium bicarbonate

   Separately dissolve the components for Solution A and Solution B, each in 300 ml distilled water. Mix Solution A and Solution B together. Make sure all ingredients are in solution, then add Sodium bicarbonate. Check to make sure that the pH is approximately 7.4, adjusting if necessary. Store in refrigerator.

2. Preparation of sheep erythrocytes:

   a. In a conical 50 ml centrifuge tube, mix 4 ml sheep blood or reconstituted erythrocytes and 16 ml HBSS.

      NOTE: If a centrifuge capable of spinning 50 ml tubes is not available, divide this into two 15 ml tubes.

   b. Centrifuge for 10 minutes, at 400 x g (usually setting 4 on a table-top clinical centrifuge).

   c. Discard supernatant and resuspend cells in 20 ml HBSS.

   d. Repeat steps b and c twice.

   e. Store washed erythrocytes in refrigerator for no more than one week.

3. Antibody coating of erythrocytes:

   a. Dilute the antiserum to sheep erythrocytes by mixing 1 ml antiserum with 19 ml HBSS.

Pre-lab Preparation - Experiment A, continued

b. Mix antiserum with washed sheep erythrocytes in conical centrifuge tube(s).

c. Incubate mixture 30 minutes at 37°C.
NOTE: The timing and temperature of this incubation are not critical.

d. Centrifuge mixture 10 minutes at 400 x g.

e. Discard supernatant and resuspend cells in 40 ml HBSS.

f. Repeat steps d and e once.

g. Store coated erythrocytes in refrigerator for one week.

4. Hypotonic lysis to remove erythrocytes from spleen cells (alternate protocol):

   a. Centrifuge spleen cells (isolated as in Unit 6 Module 2) 10 minutes, 400 x g.

   b. Decant supernatant.

   c. Tap cell pellet to resuspend cells in small amount of supernatant remaining.

   d. Add 9 ml distilled water to cell pellet. After cells have been in water for 2 seconds, quickly add 1 ml of 10x concentrated HBSS or PBS and mix thoroughly with a pipet.

   NOTE: The timing of this step is important. The lymphocytes are somewhat more resistant to the hypotonic conditions than the erythrocytes, but the lymphocytes will also be lost if exposed to the water for too long.

   e. Wash the lymphocytes by centrifuging as in step a., resuspending the pellet in 1x HBSS and centrifuging again. The cell pellet should be white rather than red at this point. If there is still a lot of red in the pellet, repeat steps a-e to remove additional erythrocytes.

   f. Resuspend the cells in 1x HBSS and count.

Method - Experiment A

1. Mix 2.5 ml of lymphocyte suspension with 2.5 ml antibody-coated erythrocytes in a 17 x 10 mm tube.

2. Seal the tube tightly with a cap or with parafilm. It is very important that you are able to invert the tube without any liquid spilling out. Even if your tube has a cap, you may want to wrap parafilm around the cap to seal it.

3. Warm the tube to 37°C by placing it in the water bath for 5 minutes.

4. Remove the tube from the water bath and gently mix it for 10 minutes by inverting it back and forth. Keep your hand around the tube during mixing to keep it warm.

5. Place the tube on ice until ready to count the cells.

6. For counting the cells, stain the white blood cells with crystal violet for easier visualization as follows:

   a. Using 1 ml pipet or automatic micropipet, remove 0.1 ml from the tube containing lymphocytes and coated erythrocytes and transfer to a 12 x 75 mm tube.

   b. Dilute sample with 0.4 ml HBSS.

   c. Add 1 drop 0.5% crystal violet to sample using pasteur pipet.

   d. Let the tube sit 5 minutes at room temperature to allow the lymphocytes to take up the stain.

7. Place a drop of the stained cells on a microscope slide and place a cover slip on top of the drop.

8. Place the slide on a microscope and focus on the cells, first using the 10x objective, and then using the 20x objective. You should be able to see both naked, stained cells, and white blood cells which are surrounded by a cluster of pale, smaller cells. These clusters are the rosettes.

9. Count the number of rosettes visible in the field. Also count the total number of white blood cells visible. Record these two numbers. You should count a large enough sample to obtain a total of at least 50 white blood cells. If necessary, count more than one field and add the numbers obtained.

> NOTE: Using the 10x objective allows you to count more cells in a single field, but it may be easier to see the cells using the 20x objective. Choose whichever objective you find easiest to use to carry out the cell counting described in step 9.

Results - Experiment A

1. Draw diagrams illustrating how the two types of white blood cells, rosetted and non-rosetted, appear under the microscope.

2. Use the cell counts obtained in step 9 to calculate the percent of lymphocytes in the spleen cell population which are Fc receptor positive.

3. If the concentration of erythrocytes in the mixture is too high, they may obscure visualization of the lymphocytes and rosettes. If this seems to be a problem, dilute the sample further in HBSS.

4. Washing the erythrocytes after antibody coating is an important step, since residual excess antibody will inhibit rosette formation.

Introduction - Experiment B

Experiment B - Identification of B Lymphocytes by Detection of Cell Surface Immunoglobulin

The B lymphocytes are the cells that produce specific antibodies when activated during an immune response. While antibodies are secreted during an active immune response, most B lymphocytes contain antibody (immunoglobulin) which remains attached to the B lymphocyte surface. This immunoglobulin, which is of the IgM class in inactivated B cells, is produced when the B cells develop in the bone marrow. Before the IgM protein is produced, the genes encoding the IgM heavy and light chains undergo rearrangement to generate unique variable regions which will combine with a specific antigen. Since this rearrangement occurs differently in each developing B cell, each B cell will produce IgM antibodies with a unique antigen specificity. Binding of the specific antigen with the IgM on a particular B cell will stimulate that cell to become activated and to begin dividing as part of the immune response to the antigen.

The presence of surface IgM can be used to identify B lymphocytes. As with many cell surface proteins, specific antibodies against the IgM class of immunoglobulin can be used as a tool to mark IgM-bearing cells. These antibodies are generated by injecting purified IgM proteins into a different species of animal and isolating the antibodies from the animal's serum. Usually antibodies which react with constant regions of IgM (i.e. determinants that are found on all IgM molecules, regardless of antigen specificity) are selected. The antibodies produced in this manner will bind to the IgM present on most B lymphocyte surfaces. If these antibodies are labeled, such as with a fluorescent dye or with an enzyme that will produce a color when it acts on an appropriate substrate, the B lymphocytes can be distinguished microscopically from the rest of the cells in a population. The IgM antibodies may be labeled directly or, alternatively, a stronger signal can be achieved by using unlabeled antibody against the IgM (called the primary antibody) in the first step followed by labeled antibody produced against the primary antibodies (called the secondary antibody) in a second step. This indirect labeling method is used frequently in clinical and research laboratories to detect the presence of many types of cell surface proteins on subpopulations of cells. This detection system can be used to determine the percent of cells bearing the cell surface protein, as well as to isolate or deplete that cell population.

Instructions are provided for both fluorescence and enzyme labeled second antibody. The instructor will indicate which label will be used. Be sure to follow the appropriate directions for the label being used. The use of a fluorescent label requires the availability of a fluorescence microscope, while the enzyme labeled antibody

Introduction - Experiment B, continued

can be visualized with a standard light microscope. Note that this procedure is written for detecting IgM on B lymphocytes isolated from mouse spleen. However, the instructor may choose to use cells isolated from a different animal, or cultured cells. If so, additional instructions and corrections to the protocol will be provided.

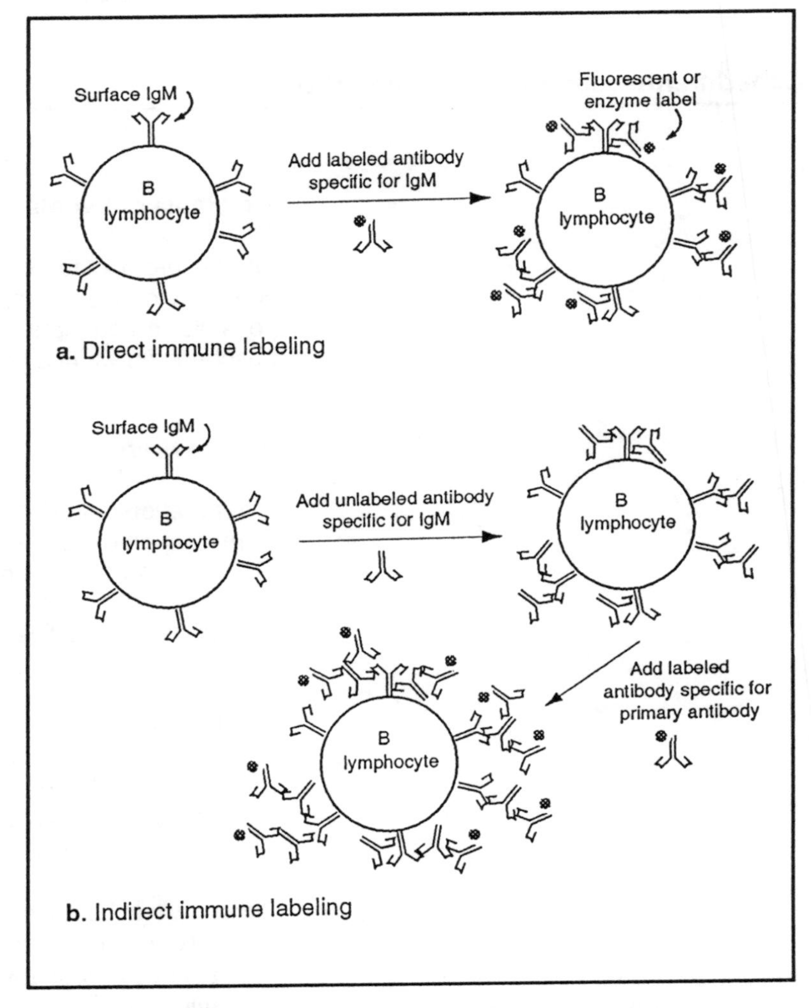

Figure 3. Direct and Indirect Immune Labeling of Cell Surface Immunoglobulin

Safety and Waste Disposal - Experiment B

1. General laboratory safety guidelines should be followed throughout the procedure.
2. If using the fluorescence microscope, the eyes should be protected from the ultraviolet light used to excite the fluorescent dye. Most fluorescent microscopes are equipped with protective shields that are positioned so that the eyes of the person looking into the objectives are protected.

Experimental Outline - Experiment B

Timetable of Events - Experiment B

The lab protocol for indirect labeling, through cell observation and counting, can be completed in a 3-hour lab period. It may be possible to complete this in a 2-hour lab period if there are no unscheduled waiting times, etc.

Pre-lab Prep

Preparation of PBS-BSA	10 min. (25 min if making PBS)
Preparation of substrate	10 min.
Washing and counting cells	20 min. (60-75 min if starting with spleens)
Dilution of antibodies	15 min. (2-3 hours if testing antibody dilutions)

Lab Procedure

Primary antibody treatment:
Mix cells and antibody against IgM and wash away unbound antibody - about 45 min.

Secondary antibody treatment:
Mix primary antibody treated cells with labeled secondary antibody, wash away unbound secondary antibody - about 45 min.

Observation and counting of labeled cells:
Calculate percent of IgM bearing cells - about 15 min.

Materials - Experiment B

1. If the use of mice is not appropriate for this lab, there are alternative sources of lymphocytes. Spleen from several different species can also be used, since antibodies against IgM are available (from Sigma Chemical Co.; see Resources). If a nearby research lab will be using other parts of an animal you might request the spleen. Bovine spleen may be available from a local slaughterhouse. Alternatively, cultured lymphocytic cell lines are available from ATCC or from research labs. A mixture of cultured IgM-positive, B lymphocytic cells and IgM-negative cells (such as a T lymphocytic cell line,) mixed in approximately 50:50 proportions, would produce results similar to splenic lymphocytes. Wash cultured cells twice with HBSS, count, and adjust concentration to 1×10^6/ml. The protocol described here assumes the use of mouse spleen cells, but the procedures would be the same for other species; just substitute antibodies with appropriate specificities.

2. If a fluorescent microscope is available, the fluorescence detection method is a useful lab technique for students to experience, and it generally yields stronger detection signals than the enzyme label. However, unless multiple fluorescence microscopes are available, students will have to wait in line to observe their stained cells. The stained cells can be stored, covered with foil, in the refrigerator for a few days for later observation.

3. A direct protocol may be used for fluorescence detection, using FITC-labeled rabbit antibody to mouse IgM. This will eliminate one incubation and washing step. However, the fluorescence obtained is generally weaker and more difficult to observe. Direct staining is also possible with enzyme-labeled antibodies against mouse IgM, but this may not yield detectable signals.

4. The success of this protocol depends critically on using the appropriate dilution of both primary and secondary antibodies. The manufacturer generally recommends a working dilution. However, it is best to determine the optimum dilution empirically for each antibody. This can be worked into the lab exercise, by having each group use a different dilution, and having all groups examine the results from the most successful group(s). Alternatively, if enough supplies are available each group can prepare multiple samples using a range of dilutions encompassing the recommended dilution.

Materials - Experiment B, continued

Phosphate-buffered saline with 0.1% bovine serum albumin (PBS-BSA) 100 ml

Lymphocytes isolated from mouse spleen and purified by acetic acid lysis or hypotonic lysis (see Unit 6, Module 22 - also see lab preparation below), 2 x 10^6/ml in PBS-BSA (2 x 10^7 cells total)

12 x 75 mm centrifuge tubes, 10

Microscope slides, 20-30

Cover slips, 20-30

Pasteur pipets, 30-40

Bulbs, 10

Goat antibodies against mouse IgM, diluted appropriately in PBS-BSA (see Note #4 above), 2 ml

1 ml pipets, 30-40

Pipet pumps, 10

Gloves

Ice

90% glycerol in PBS-BSA

For fluorescence detection:

Fluorescein isothiocyanate (FITC)-labeled rabbit antibodies against goat immunoglobulin, diluted appropriately in PBS-BSA, 2 ml

For enzyme-linked color detection:

Alkaline phosphatase-labeled rabbit antibodies to goat immunoglobulin, diluted appropriately in PBS-BSA, 2 ml

Alkaline phosphatase substrate (nitroblue tetrazolium (NBT) and 5-bromo-4-chloro-3-indolylphosphate (BCIP))

For fluorescent detection:

Fluorescein isothiocyanate (FITC) labeled goat antibody against rabbit immunoglobulin (secondary antibody)

For enzyme linked color detection:

Alkaline phosphatase-labeled goat antibody against rabbit immunoglobulin (secondary antibody)

Alkaline phosphatase substrate (nitroblue tetrazolium (NBT) and 5-bromo-4-chloro-3-indolylphosphate (BCIP)), made fresh by adding 6.5 µl BCIP and 33 µl NBT to 5 ml PBS-BSA

Pre-lab Preparation - Experiment B

1. Preparation of PBS-BSA: To 100 ml PBS, add 0.1 g bovine serum albumin, stir to mix.

   PBS can be purchased as a 1x or 10x solution or as a powder. If prepared in the laboratory, dissolve the following in 150 ml distilled water:

   2.19 g NaCl
   1.28 g KH_2PO_4
   2.63 g $Na_2HPO4 \cdot 7H_2O$

   Adjust pH to 7.2, add distilled water to bring volume to 250 ml.

2. Hypotonic lysis to remove erythrocytes from spleen cells:

   a. Centrifuge spleen cells (isolated as in Unit 6 Module 2) 10 minutes, 400 x g.

   b. Decant supernatant.

   c. Tap cell pellet to resuspend cells in small amount of supernatant remaining.

   d. Have ready a pipet or syringe containing 1 ml of 10x concentrated HBSS or PBS.

   e. Add 9 ml of distilled water to the cell pellet. After cells have been in water for 2 seconds, quickly add 1 ml of 10x concentrated HBSS (see lab preparation for Experiment A) or PBS and mix thoroughly with a pipet.

      NOTE: The timing of this step is important. The lymphocytes are somewhat more resistant to the hypotonic conditions than the erythrocytes, but the lymphocytes will also be lost if exposed to the water for too long.

   f. Wash the lymphocytes by centrifuging as in step a., resuspending the pellet in PBS-BSA, and centrifuging again. The cell pellet should be white rather than red at this point. If there is still a lot of red in the pellet, repeat steps a-f to remove additional erythrocytes.

   g. Resuspend the cells in PBS-BSA and count. Adjust the cell concentration to 2×10^6/ml.

3. Dilution of primary and secondary antibodies: See Note #4 above regarding dilutions. If desired, the instructor can prepare a series of dilutions and test them prior to the lab, and use the optimal dilution in the lab. Antibodies should be diluted

Pre-lab Preparation - Experiment B, continued

fresh, no more than 12 hours before the lab. If desired, aliquot diluted antibodies into labeled tubes for groups. FITC-labeled antibodies should be protected from light as much as possible.

4. Preparation of cells: Cells should be washed once in PBS-BSA by centrifuging 10 minutes at 250 x g, resuspending in PBS-BSA, and centrifuging again. Resuspend cell pellet in PBS-BSA so that cells are at a concentration of 2 x 10^6/ml.

5. For alkaline phosphatase label-preparation of substrate: Both substrate components are available as tablets from Sigma Chemical Co., which are dissolved in the indicated amount of distilled water just before use. Make substrate solutions as fresh as possible, and protect from light until used.

Method - Experiment B

NOTE: It is important to keep the cells and reagent on ice throughout the procedure. If the cells become warm, the IgM-antibody complexes may be internalized and will be difficult to visualize. The addition of 0.2% sodium azide to the PBS-BSA would inhibit internalization, but since azide is highly toxic it is not recommended for use in a classroom laboratory.

1. Transfer 1 ml of lymphocyte suspension to a 12 x 75 mm centrifuge tube.

2. Centrifuge 10 minutes at 250 x g.

3. Carefully remove supernatant with pasteur pipet and discard.

4. Add 0.1 ml diluted antibody against IgM and gently resuspend the cell pellet by tapping the bottom of the tube or by using the pipet to break up the pellet.

5. Incubate on ice 15 minutes.

6. Add 1 ml PBS-BSA. Mix.

7. Centrifuge 10 minutes, 250 x g.

8. Remove supernatant and discard.

9. Add 1 ml PBS-BSA and resuspend the cell pellet.

10. Repeat steps 7 and 8.

Method - Experiment B, continued

11. For fluorescent detection (see below for enzyme linked color detection):

    a. Add 0.1 ml FITC-labeled goat antibody against rabbit immunoglobulin and resuspend the cell pellet.

    b. Incubate 20 minutes on ice.

    c. Wash cells as in steps 6-10 above.

    d. Add 0.1 ml PBS-BSA and resuspend the cell pellet.

    For enzyme linked color detection:

    a. Add 0.1 ml goat antibody against rabbit immunoglobulin and resuspend the cell pellet.

    b. Incubate 20 minutes on ice.

    c. Wash cells as in steps 6-10 above.

    d. Add 0.1 ml peroxidase substrate and resuspend the cell pellet.

12. Place a drop of 90% glycerol on a microscope slide. Add a drop of stained cell suspension. Place a cover slip over the cells, taking care to avoid air bubbles.

13. For fluorescence detection:

    a. Observe cells using microscope equipped with ultraviolet light source and filters for fluorescence detection Use a 20x or 40x objective.

    b. First count the total number of lymphocytes using normal light. The lymphocytes will be small, round cells with little cytoplasm.

    c. Then, in the same field, use the ultraviolet light to count the number of fluorescent cells. The fluorescence should have a "patchy" appearance on the lymphocyte surface.

    For enzyme linked color detection:

    a. Observe cells using a 20x or 40x objective.

    b. Count the total number of lymphocytes in the field. The lymphocytes are small, round cells with little cytoplasm.

    c. Count the total number of lymphocytes which have stained with peroxidase. These cells will have a purple stain over the surface.

Results - Experiment B

1. Use the cell counts obtained to calculate the percent of B lymphocytes in the lymphocyte sample.

2. Is the percent of B lymphocytes calculated from your data consistent for what one would expect to see in this cell sample? Explain.

3. The critical parameter is antibody dilution. No positive cells may indicate that one of the antibodies is too dilute. A high degree of background staining (resulting in no distinction between positive and negative cells) may indicate the antibodies are too concentrated and nonspecific binding is occurring.

4. You may want to include a control sample which is treated with the secondary antibody, but not the primary antibody. This may help determine the degree of nonspecific binding of the secondary antibody.

5. Another way to reduce nonspecific staining is to add one more washing step between each antibody addition. This will increase the time required for the exercise.